WHERE TO FIND GOLD IN THE DESERT

JAMES KLEIN

Gem Guides Book Company
315 Cloverleaf Drive, Suite F
Baldwin Park, CA 91706

Revised Edition Copyright © 1994
By James Klein

Cover by Marios Savvides

All rights reserved. This book, or any portion thereof, may not be reproduced in any form, except for review purposes, without the written consent of the publisher.

Printed in the United States of America.

Library of Congress Catalog Card Number: 95-75456

ISBN 0-935182-81-0

Contents

WHERE TO FIND GOLD

California Areas

1. Rosamond-Mojave Area 8
2. El Paso Mountains 16
3. Randsburg Area 24
4. Barstow Area 34
5. Dale Area 44
6. Anza-Borrego Area 52
7. Chocolate Mountains-Tumco-Potholes Area 64
8. Other California Areas 73

Arizona Areas 80
New Mexico Areas 92
Utah Areas 98
Southern Nevada Areas 100

HOW TO FIND GOLD

Geology of Placer Deposits 105
Some Tips on Desert Prospecting 113
Panning in the Desert 122
How to Stake a Claim 126
Glossary 130

Pacific Gold-Copper Mine, Stedman District. This photo of underground workings in the mine, in San Bernardino County, was taken in the early 1900s.
(Courtesy of California Department of Mines and Geology)

Introduction

The Southern California desert contains one of the largest reserves of gold remaining to be discovered. In this new edition, you will find more information on dry washing and metal detecting, as well as more gold bearing locations in the desert regions of the southwestern states.

The tales told and the bonanzas known of the desert fill you with this kind of excitement. There are more stories told of lost treasure in the desert than any other area. It is not hard to believe these tales when you become familiar with the harsh beauty and isolation of the desert, and with the vast amounts of wealth already taken from it.

I have my own lost desert treasure tale I can tell. When I first began prospecting, there was a wash where I found quite a bit of placer gold. The wash had a drain-off where the gold was found. I had worked at the foot of the drain; later I realized that the heavier nuggets would be at the head of the drain where the material would back up before getting out of the narrow drain-off. Although I have returned to the area many times, I have never been able to find the exact spot again. Maybe you will find it.

In this book, you will learn where the gold has been found in the past, where it can be found today, the lost treasure tales of each area, and points of interest.

Some words of warning to you before you set out. Always be sure your vehicle is in good condition when traveling in the desert. It seems that every summer you read where someone's vehicle breaks down on some back road and its occupants are found dead from exposure.

Be sure that you have plenty of gas and an extra five-gallon can of water for both you and your vehicle. I always carry an extra fan belt and tools. Always let someone know where you are going and when you plan to return.

Having the eternal optimism of the prospector, I expect that I'll discover a spot where wind has exposed the desert floor and find it covered with gold nuggets. Now, let's go find some of that gold!

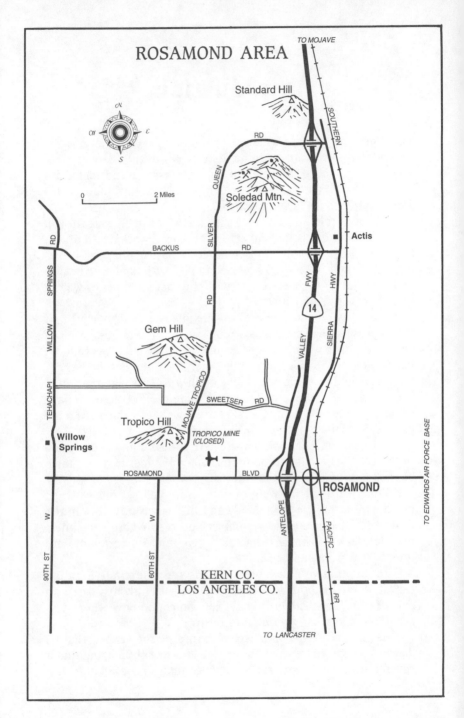

California Areas
1. Rosamond-Mojave Area

This district is the closest to Los Angeles. Shown as the Mojave-Rosamond district by the State of California Department of Mines and Geology, it sits off Highway 14 between the towns of Lancaster and Mojave. The hills are studded with old mine shafts and tunnels. Abandoned shacks and headframes dot the landscape. It's like a ghost town of mines. Some of the claims are still valid, and there is a caretaker who will chase you off if you're trespassing on one of them.

The first mining here was for fire clay and began in the 1870s. Gold was discovered on Standard Hill in 1894 by a man named Bowers, and on Tropico Hill, then known as Hamilton Hill, by Ezra Hamilton in the same year.

How Hamilton made his find is interesting. He owned a pottery plant in Los Angeles and one day was panning some fire clay from the hill when he found gold. He rushed to the site, located the lode, and became a very rich man. One wonders what gave him the notion to pan the clay. He may have seen some clue while working with the clay. I would like to have all the pipe and bricks made with the clay for over ten years before the discovery. Los Angeles is, in truth, a city paved with bricks of gold.

Mining boomed in the area for about twenty-five years and then went quiet until 1931, when the depression made all the gold districts active again. This lasted until the war closed all the mines in 1941. Since the war, there has been minor activity at some of the mines. There is much to be seen here; it's well worth a trip.

From Los Angeles take the Golden State Freeway (Interstate 5) to the Antelope Valley Freeway (14); go past Lancaster (on 14) to the Rosamond turnoff. Go left on Rosamond Boulevard and follow the signs to the Tropico Mine and Gold Camp; this road passes the mining areas and Gem Hill and takes you back to Highway 14.

An aerial view of the Tropico Mine and Ghost Camp. (Courtesy of Burton's Tropico Gold Camp)

Exposed treasure mine, Mojave District.
The photo shows the Kern County mine as it looked about 1914.
(Courtesy California Department of Mines and Geology)

Points of Interest

At one time, the Tropico Mine and Gold Camp here had a museum and gold tour to visit, but it has been closed for some time. The Gold Camp was made up of buildings from the area that were moved here to be saved from destruction by time and the elements. The Mining Days Museum contained some of the prettiest gold ore samples you'll ever want to see. They even had a big gold panning contest and treasure hunt here every spring.

Several prominent Californians have been involved with mining here. Former Governor Goodwin Knight and his father operated the Elephant Eagle Mine in the 1930s. Governor Knight held the claim on it right up to his death. Harvey Mudd, who has a college named after him, was involved in the Cactus Queen Mine north of Willow Springs. It produced six million dollars in gold and silver in one six-year span. I guess that means if you strike it rich here, you might go on to be Governor or maybe even get a school named after you!

Gem Hill is one of the most popular collecting areas for rockhounds in the desert. You'll find petrified wood, jasper, agate and some opal here.

Prospecting Tips

The mines in this area have produced over $23 million in gold and silver. The gold occurs in quartz veins up to forty feet thick. There was some rich ore found on the surface during the early days, which led to discovery of the lodes. I have never heard of any placer work being done here, but it stands to reason that there must be some placer deposits nearby with such rich surface float. The next large lode may be sitting a foot below the surface waiting to be discovered.

Treasure Tales

On a butte near Mojave there is said to be a treasure dating from the early days of the Spanish in California. The butte is off the old trail, so the party that left it either was lost or was chased there by Indians. Spanish armor and other artifacts, along with

Woody Woodruff displays his nine-and-a-half-ounce "Mojave Camel" nugget, found while nugget shooting. (Courtesy California Prospecting Co.)

Spanish coins dating from the 1600s, have been found scattered around the base of the butte. They must have used the butte as a landmark; it's right off the old Borax Road outside of Mojave.

Palmdale Treasure

When the Governor Mine near Palmdale was producing gold, the owner decided to hoard some of it rather than convert it into cash. An employee knew of the hoard, which was kept in the company safe. He felt that the owner was keeping it illegally and that the owner would not report it if it was stolen. One night he broke into the safe and took the gold and some money that was with it. He buried the gold nearby, figuring he would dig it up later when he could leave the camp without suspicion. The owner did call the police, and the money was found in the man's belongings. The man was convicted of the crime and was sent to jail. He refused to reveal where he had buried the gold.

Years later, when he got out of jail, he returned to the area. He was watched constantly by the police. Unable to recover his loot, he left Palmdale. He later went to jail again. This time he became ill and was sent to the prison hospital where he told an orderly where the gold was buried and drew a crude map. The orderly was never able to find it. He gave the map to another man, and when this man tried to locate it with a metal detector, he was shot at and never returned. Somewhere nearby is $150,000 in gold waiting to be found.

Historical Note

Near the Tropico Mine is the old stage stop of Willow Springs. The State of California Historical Landmarks Monument reads: "Visited by Padre Garces in 1776 while following old Horse Thief Trail, later known as Joe Walker Trail. Fremont stopped here in 1844. The famished Jayhawk party of 1850 found water here while struggling from Death Valley to Los Angeles. Still later this was the stage station on the Los Angeles-Havilah and Inyo Stage Lines."

Here is a perfect example of the fickle finger of fate. Those hundreds of pioneer travelers walked over millions of dollars in gold on their way to the water at the spring.

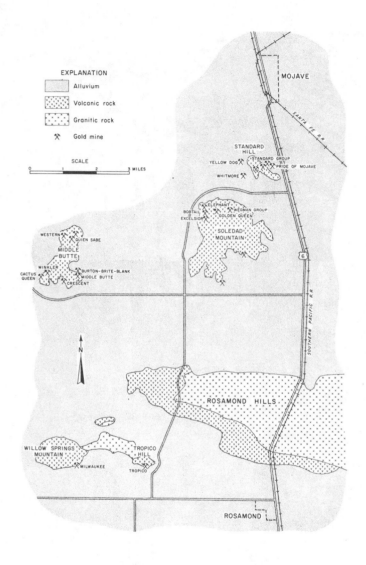

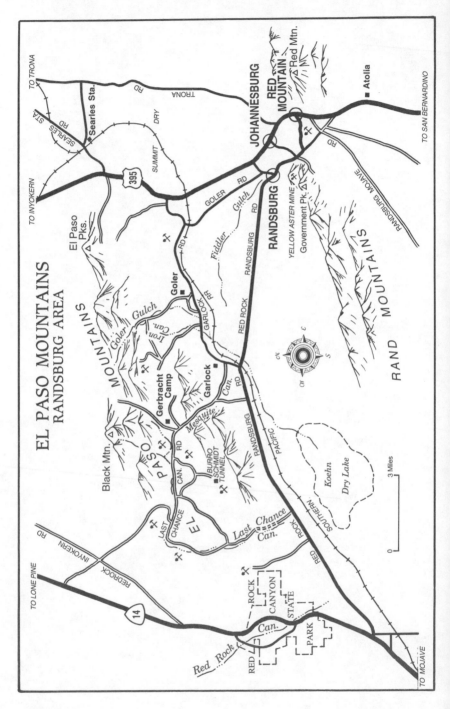

2. El Paso Mountains Area

The El Paso Mountains are not considered a part of the Mojave Desert gold district by the State of California Division of Mines and Geology. They are placed in Basin Range District. Most prospectors still think of them as a part of the desert. The El Paso Mountains have it all, natural and man made wonders, opal mines and gem stones, gold and silver, lost mines and ghost towns.

Red Rock Canyon sits on the western tip of the El Paso Mountains and is now a State Park. You may be able to camp here and use it as a base for your prospecting trips into the mountains.

Up in Goler Gulch there has been some unique mining going on. Someone was using a skip-loader to scoop up gold-bearing gravels and load them into a large tractor trailer. The gravels were apparently taken to Bakersfield where the dirt was washed. No one was there at the times I've been there, so I don't know how successful the operation has been. One thing I know is that people don't spend that kind of money unless they're getting something back. Benson Gulch, which runs off of Goler Gulch, has been the most productive gold-bearing area.

The ghost town of Garlock lies between Highway 14 and Goler Gulch. The State of California Historical Landmark reads: "SITE OF THE TOWN OF GARLOCK." In 1896 Eugene Garlock constructed a stamp mill near this spot for the crushing of gold ore from the Yellow Aster Mine on Rand Mountain. Known originally as "Cow Wells" by prospectors and freighters during the 1880s and early 1890s, the town of Garlock continued to thrive until water was piped from here to Randsburg in 1895 and the Kramer-Randsburg rail line was completed the same year. Garlock still has some residents today. Most of the old buildings are on private land and are fenced off.

*Old buildings at the ghost town of Garlock.
(Courtesy of M. Broman)*

I heard a funny story from one of the old timers there. A while back a road crew was working near Goler Gulch when two schoolmarms from back east drove up and asked them where they should look for a gold nugget to take back as a souvenir of their trip. Upon hearing this humorous request, some crewmen sent the elderly ladies over to one of the old mines and told them to look in the tailings. The ladies returned about a half-hour later. They thanked the men for their help and showed the flabbergasted crew a slab of gold the size of your palm that they had picked right off the top of the tailings. *You never know.*

Summit Dry Diggens is east of the main body of the El Pasos. There is a great deal of activity there on the weekends. There are many active claims here, so get permission if you're working on posted land. This was also known as "Poor Man's Diggens" because the gravels are so low-grade. However, you can still find a little color. It was from here that three miners, disgusted at the lean pickings, wandered down to the Rand Mountains and made the discovery that led to the boom at Randsburg.

From Los Angeles, take the Golden State Freeway (Interstate 5) to the Antelope Valley Freeway (14). Take Highway 14 all the way to the Red Rock-Randsburg Road. You will pass through the towns of Lancaster and Mojave. The Red Rock-Randsburg Road runs off to your right from its junction with Highway 14 at a spot known as Wagon Wheel. There is a good-sized, deserted building at the junction.

Goler Gulch is past the ghost town of Garlock. The road forks right before you get to Garlock; take the road going left. It continues on until it hits Highway 395. It is three miles from the fork to Iron Canyon Road. This road will take you into Goler Gulch. This road also forks. Take the road to the right, which will take you past the Goler Cemetery and through Benson Gulch.

All the roads off the main road are dirt, but in good traveling condition. About a half-mile past the Iron Canyon Road is Goler Road. The Last Chance Canyon Road is six-and-a-half miles from Highway 14 on the Red Rock-Randsburg Road and takes you back into the El Paso Mountains. There has been quite a bit of placer gold taken out of here. It takes you past the old mining camp of Cudahay, now a ghost town, Copper Basin, and to Burro Schmidt's Tunnel.

Continuing on Highway 14 past the Red Rock-Randsburg Road will bring you into the Red Rock Canyon Area. This is now a State Park, and prospecting is not permitted. There were some fairly large nuggets found here in the old days. I don't suppose, if you found a big nugget now, anyone would say anything.

About seven-and-a-half miles from the Red Rock-Randsburg Road on Highway 14 is a dirt side road on your right, that will take you to an opal mine. This is a private claim. The owner will let you search for the opals for a small fee. He showed me some beautiful stones and claims it's better than gold.

Another seven-tenths of a mile farther on Highway 14 you will come to the Red Rock-Inyokern Road. Take the Red Rock-Inyokern Road 4.6 miles to Hart's Place Road. Turn right into Last Chance Canyon. This is a better road than the one coming up from the Red Rock-Randsburg Road.

Prospecting Tips

Most of the gold in the El Paso Mountains has been placer gold, and is found today in gravel benches on canyon walls. The early miners were able to recover it from the washes where it had washed down from the gravel benches. You will have to go to bedrock in the washes now to mine it. There have been some quartz veins found and worked with some success. The gold-quartz veins occur in granite and schist. There are petrified wood, palm, jasper and agate, as well as opal to be found in the El Paso Mountains.

Treasure Tales

By far the most important treasure tale of the El Paso Mountains is the Lost Goler (or Goller) Mine. There are several versions of the tale. One thing that confuses the stories is that there were two men named Goler who prospected in the region and made rich discoveries that were lost. The version I like best is as follows:

A prospector named Goller had been prospecting in the mountains farther east. He was returning to San Bernardino for supplies and stopped for water from a spring running in one of the canyons running out of the South side of the El Paso Mountains. When he knelt down to drink, he came face to face with the most beautiful gold nuggets he had ever seen. He claimed the stream bed was full of them. Lacking the supplies to stay and mine the gold, he picked up as much as he could and headed for San Bernardino. He figured he would pick up enough provisions to last several months and return and mine the gold. Along the way he stuck his rifle in the ground to use as a marker to guide him back to the rich placer. He got to San Bernardino all right, but he would never find the gold again.

After he got to San Bernardino he decided to celebrate his good fortune by downing a few with the boys at one of the local bistros. Mr. Goller proceeded to go on a monumental drunk. The drinking spree lasted for several weeks. When he had sobered up, he prepared to return to his find. Just as he started to leave, the area was hit with one of the worst storms in years. When at last able to set out, he found the storm had changed all his landmarks. This, along with his drinking bout, befuddled his memory so much he was never able to locate the little canyon again. He wandered the hills for years and then disappeared.

There is no doubt that Goller found a rich placer deposit in the El Pasos. Too many men saw the gold to deny that fact. It must have been in the El Paso Mountains because it was there that Goller searched all those years.

One last story to note. Years later a rancher in the area found a rusted rifle like the one Goller carried, stuck into the ground on a mound near Red Rock Canyon.

Lost Treasure of Robbers' Roost

At one time three lost treasures were here, but now there is only one. The two treasures recovered came to over $100,000. The two treasures that have been found were both bandit loot. The one we are interested in is the result of an angry Mother Nature. Robbers' Roost is near Freeman Junction, which is at the intersection of Highway 14 and Highway 178 several miles up Highway 14 from Red Rock Canyon. Highway 178 is known as Walker Pass (a State Historical Landmark) and takes you into the Lake Isabella recreation area.

At one time there was a stage station located in a wash north of Robbers' Roost a half mile from Freeman Junction. One night a stage coach with a strong box holding $25,000 in gold stopped for relief there. The passengers had gotten out for a rest, and the stage was left standing in the wash. A storm was brewing and a small amount of water was running through the wash. All seemed to go quiet for a moment, then a sound like thunder came rolling down the wash. The startled people looked up to see a giant wave of water 15 feet high carrying immense boulders like giant corks hurling at them. No one had time to flee. Men, horses, buildings, stage coach and gold were all buried beneath the debris carried by the water.

Only one man survived the deluge; he was a worker who had been sleeping in a shelter on higher ground when it struck. He himself lost $300,000 when the water washed the station away. Over the years since that tragic night the wind and water must have washed a lot of the overburden away. Some parts of the stage coach and a rusted pistol have already been found.

Iron Canyon Treasure

When Garlock was still alive, an old Indian used to bring in some nuggets to trade at the general store. Some men tried to follow him and learn the source of his gold. They were able to trail him as far as Iron Canyon where they lost him; he must have been working one of the smaller canyons leading into Iron Canyon.

Point of Interest

Be sure to visit Burro Schmidt's tunnel. William H. Schmidt was his real name. He spent nearly 30 years digging the 2000-foot-long tunnel. It was even featured in *Ripley's Believe It or Not* column. Some people think that the tunnel was just some crazy old man's stunt. A friend who teaches mineralogy tells me that old Burro Schmidt was crazy like a fox. He got a look at the ore that came from the tunnel. He says that's where the money came for him to afford to spend 30 years building a tunnel to nowhere.

There is a lost treasure tale here. Somewhere along the walls is said to be an entrance to a room where there is a wash tub half-filled with gold. The entrance is supposed to be covered with rubble.

End of a shift. A shift, circa 1900, ascends after a work tour in a mine. Many of these miners were Cornishmen, "Cousin Jacks." (Courtesy of California Division of Mines and Geology)

The town of Randsburg. (Courtesy of M. Broman)

3. Randsburg Area

If you don't have gold fever yet, a visit to Randsburg will give it to you. Even if you're an old timer, Randsburg will excite you. To me, this area has more mining atmosphere than most of the Mother Lode country. I think a lot of it comes from the residents still remaining here. They all are confident that tomorrow will bring the shout of "Eureka" and the hills will resound again with the clank of machinery and the pounding of the stamp mills. They may be right; three times in the past the area seemed doomed, and three times the area has boomed. First it was gold, then scheelite or tungsten, and then silver that brought prospectors, miners, gamblers, adventurers and ladies of the evening rushing to these lonely hills.

Gold was first discovered in 1895 by three prospectors from Summit Dry Diggens. They had exhausted their provisions and had given up the idea of ever getting rich at the "poor man's diggens." One of the three had found a little color on Rand Mountain once, and they thought they would try to find the source. They only had a slab of bacon and a few beans left when they arrived at Rand Mountain.

They had been grubstaked by the wife of one of the men, a woman doctor by the name of Rose La Monte Burcham of San Bernardino. She had spent her last penny on some pipe dream, she had told them.

It was three pretty discouraged prospectors who sat down to rest in a gully on Rand Mountain. They had been searching for three days with no luck at all. One of the three, John Singleton, an ex-carpenter, began to chip away at the ledge he was sitting on. Something about the rocks caught his eye. He picked up one of the pieces and studied it. All of a sudden he sprang to his feet and began dancing around. "GOLD, GOLD, GOLD by damn it's GOLD," he yelled. This is how the richest mine in the area was discovered. The Yellow Aster produced over 12 million

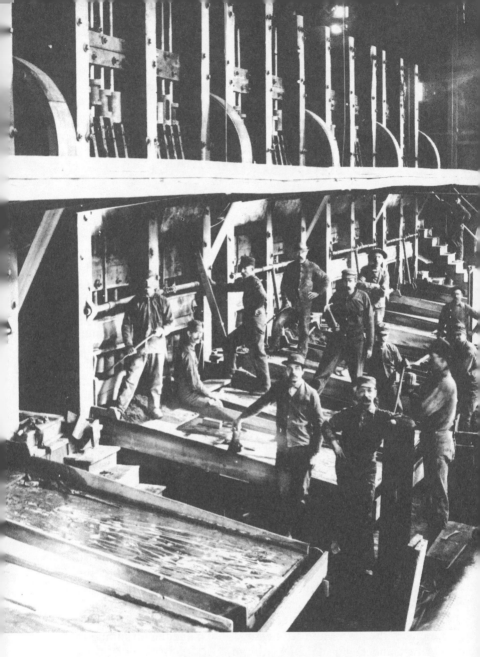

Stamp mill at the Yellow Aster Mine.
(Courtesy Title Insurance and Trust Company)

dollars in gold. The three prospectors who were broke, hungry and ready to return to regular jobs became millionaires. The first gold mined by the three was done by hand and amounted to $800.00 for the first three months. Later the Yellow Aster would produce as much as $63,000.00 in one month alone.

Prospectors swarmed to the area and mines sprang up over the hills. The boom lasted until 1905 when the mines began to play out. During the gold mining days, the miners were plagued with a heavy gray material that they found with the gold. They cussed at it and threw it out. One day one of the men decided to have it assayed to see what it was. It wasn't worthless, but it wasn't worth much, either. It was scheelite, the base ore of tungsten. A man could get $3.00 for twenty pounds of it. Since the ore was so plentiful, in some places it covered the ground with pieces so large they called them potato patches, some of the men were able to make bean money by just picking up the pieces off the ground.

The war in Europe caused the price to climb. (Tungsten is used as an alloy to strengthen steel and makes lighter, stronger weapons.) By 1910 the price of scheelite had risen to $100.00 a unit. Again the Rand District boomed. People came from all over to pick the hills clean of the precious ore.

The main lode was discovered by two prospectors about five miles from Randsburg. The men were named Atkins and De Golia; a combination of their names was given to the strike, Atolia. By 1915 there were two thousand men working claims around Atolia. This boom ended in 1918 when the war ended. Again the district didn't die, it just went to sleep for awhile.

The next boom was not long in coming. In June of 1919 two prospectors grubstaked by the former Sheriff of Kern County, John Kelley, made the discovery that would bring the dreamers and the schemers rushing to these hills once again. This time it was silver that brought them. How the thousands of prospectors and miners could have walked over the silver veins all those years without locating them no one knows. It makes you wonder how much more still remains to be discovered in the other mining districts. One thing it does teach you is that you shouldn't concentrate so hard on finding the tree, that you don't see the forest. The Randsburg District is sleeping again. Who will wake the slumbering giant next? Maybe it's your turn.

There are two ways to get to this region. If you're coming from the north part of Los Angeles, take the Golden State Freeway (5) to the Antelope Valley Freeway (14). Take the Antelope Valley Freeway all the way to Mojave. Take Highway 14 out of Mojave to the Red Rock-Randsburg Road. Follow it all the way to Randsburg. There are road signs to direct you.

From Los Angeles, take any of the freeways to the city of San Bernardino. Take Highway 395 north out of San Bernardino. Highway 395 is combined with Highway 15 here. When you get over the mountains, it branches off. Be sure that you take the Highway 395 turn off. It comes before you get to Victorville. Highway 395 takes you past Atolia, Red Mountain, and Johannesburg. The road to Randsburg is about a mile past Johannesburg.

Prospecting Tips

There are both lode and placer deposits in this district. The main production has come from the lode deposits. Most of the placer gold has come from the dry placers in the Rand Mountains north of the town. The gold mostly occurs in schist (known as the Rand Schist) where the rocks have a pale red or pinkish color. I have taken gold out of a large pit that is on the outskirts of town. It is in the walls of the pit along the lines of what was once bedrock.

Two other designated gold districts are nearby. Spangler, about ten miles northeast of Johannesburg, had one large producing mine, the Spangler. The gold occurs in quartz here and is said to have contained as much as an ounce of gold per ton. The Rademacher district is about five miles beyond the Spangler district in the direction of Ridgecrest. The gold occurs in quartz veins here, too.

To get to the Spangler district, take Highway 178 east from its junction with Highway 395 a few miles north of the Randsburg Road. The mines are in the Spangler Hills to your right after you pass Teagle Wash. The mountains and valleys to the east have yielded many Indian artifacts such as arrowheads and potsherds. There is also agate, jasper, and chalcedony to be found there.

Historical Note

Indian petroglyphs may be seen in the Lava Mountains

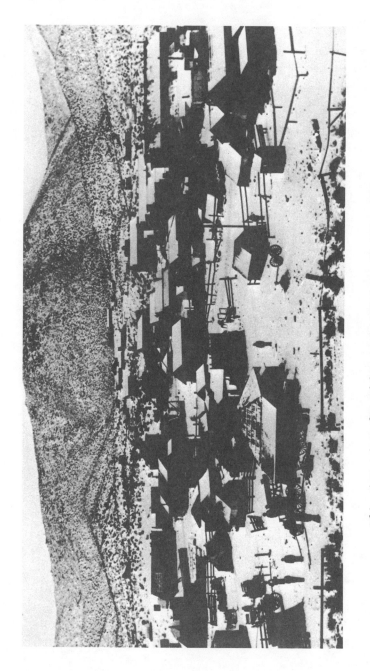

This winter view of Randsburg, taken in the early 1900s, looks east. (Courtesy of California Division of Mines and Geology)

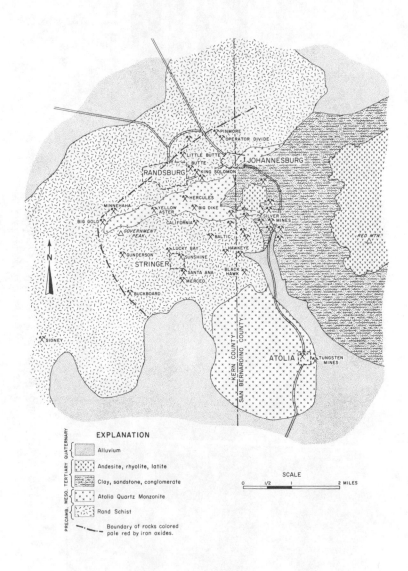

nearby. Take the Trona Road midway between Red Mountain and Johannesburg to the Steam Wells Road. Go three miles on the Steam Wells Road until you come to a road on your left. Take the fork left to the petroglyphs. If you have never seen Indian petroglyphs, you will be amazed at the amount of work and time they had to put in to create them. The district was named for the Rand district in South Africa, one of the richest in the world. The Yellow Aster Mine was originally named the Rand Mine by its discoverer.

Treasure Tales

Most of the treasure tales are to the north and the east. I've deliberately stopped your search for desert gold at Randsburg on the north. I don't feel that people should be encouraged to enter the Death Valley region until they have learn desert savvy in less severe areas. I will touch on a few of the more prominent ones only.

The Lost Chinaman

One day when two men were loading borax at Searles Lake, they noticed a man staggering across the flat. When they went out to meet him he collapsed at their feet. One of the men was John Searles, for whom the lake is named. Shorty Harris, one of the best known desert prospectors, claimed to be the other man. Some versions say the other man was a borax worker named Parkinson. He was given a chunk of almost-pure gold that weighed nearly 15 pounds. The man, who died later from exposure, was a Chinaman who had been working in a borax mine in Death Valley. He had gotten into an argument with his employer and set out across the desert without sufficient supplies. He said that he had found the rich ore in a small canyon somewhere in the Panamint Mountains.

The Lost Goller Mine

This is the tale of another Lost Goller Mine and is sometimes confused with the Lost Goler Mine in the El Paso Mountains. This Goller mine is in a gulch in Death Valley. John Goller was one of the early forty-niners who crossed Death Valley heading for the gold fields. His party ran low on water and Goller was sent on scouting trips along the canyons of the Panamint Mountains running into Death Valley.

On one foray, he discovered a rich placer in one of the gulches and carried out a small bag of nuggets. Many people saw the nuggets and Goller became a successful businessman in Los Angeles. He was respected for his honesty, so there is no doubt that his story is true. The winds shift the desert sands constantly. I've gone to sleep there and awakened to a completely different landscape. You'd swear some giant picked your whole camp up during the night and moved it. The sands must have buried Goller's nuggets, because he was never ever able to find them again.

The Lost Breyfogle

The Lost Breyfogle is one of the most famous in California. The story begins in 1862 when three prospectors set out for Nevada where a big silver boom was in progress. They made camp near the Panamint Mountains one night. The camp was attacked by Indians and two of the prospectors were killed. The other man, Breyfogle, managed to escape. His only route to safety was across Death Valley. He had no supplies, not even a canteen of water. He found a spring but the water was polluted and it made him delirious.

Later, he could only recall that he'd seen what looked like green vegetation on the side of the mountains. He thought that there would be a spring there. It turned out to be only some mesquite.

The mesquite is your clue; it was near there that the deranged Breyfogle found the beautiful rich ore that has haunted the dreams of treasure seekers for over a century. The gold-streaked quartz clung to by Breyfogle even in his delirium is proof of his discovery.

He returned to Death Valley many times searching for the vein. He disappeared on one of those journeys, as many others have done in that desert. Most searchers place the vein in the Funeral Ranges.

The Lost Gunsight

A party of forty-niners was crossing Death Valley when four of them set off to take a short cut across the Panamint Mountains (I just love that name, the Pan-a-mints). They wandered along the base of the mountains for several days, seeking a route over the range.

One day, while the group was at the base of Tucki Mountain, Captain Towne, the leader of the party, struck his rifle barrel against a boulder, knocking off the sight. Towne made a new sight for his gun with some ore he found and later discovered it was pure silver. Most of the tales say that they knew right away what they found, but were too worried about saving their lives to do anything more than pick up as much of the ore as they could carry and get out. They claimed that the small canyon was covered with the rich ore.

The Captain and another member of the party, a James Martin, (Martin is said to be the discoverer by some) returned to Death Valley the next year to try to relocate the mine. Their bodies were found in a small canyon; they had been murdered. Some claim that the Indians killed them to keep the white men from a sacred place.

The Lost Alvard

Alvard (or Alvord) was a prospector who discovered a rich ledge of gold while searching for the Lost Gunsight Mine. He was with a party of men known as the Mormons. One day he left camp by himself and was gone for over a week. When he returned, he brought back some black ore rich in gold. The party's supplies were running out, so they decided to return to Los Angeles to re-outfit themselves and have the ore assayed. The ore turned out to be richer than any of them dreamed.

Now the story becomes confused with double-dealings and accusations. Alvard at first didn't want to cut anyone in, and then made a deal with one of the other men. Potshots were taken at Alvard and threats were made against his life. He did finally make it back to Death Valley, but he never found the ledge again.

His body was found one day by some other prospectors. He had been killed and robbed of his outfit by a man named Jackson who had lured him out on the pretense of finding a lost mine.

There are many other tales of lost mines and buried treasure to be told of this region. There is the Lost River of Gold under Kokoweef Peak, said to be worth millions, and the Lost Jayhawker Cache of gold and silver coins buried under a chalk-white cliff in the valley.

You can search for coins hidden by the Indians after they had ambushed and killed a wealthy Mexican family and their party, or Ramies Lost Ledge, and then there's the . . .

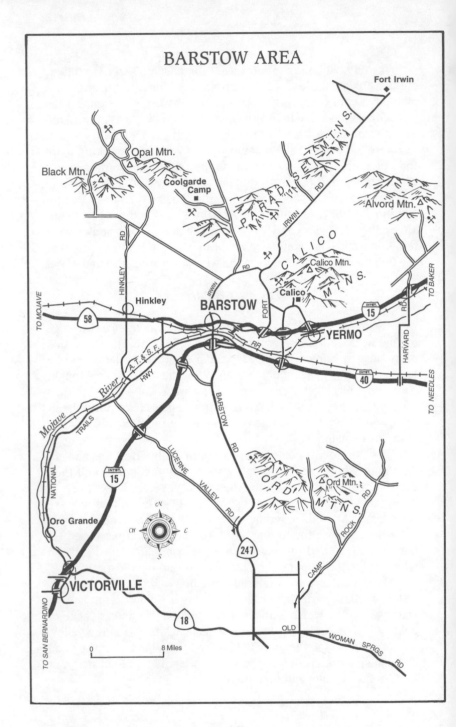

4. Barstow Area

Barstow itself is not a mining district. We will use it as a jumping-off spot to several smaller gold districts such as Oro Grande, Ord and others. Barstow was originally a supply center and stopping-off place along the old Mojave River trail.

I recently had an interesting proposal from a woman who owns quite a bit of property there. Several years ago, she purchased a piece of property outside of town that had an old mine on it. She didn't think anything about it as she had no knowledge of mines or mining. When the price of gold went up, she got to wondering if there might not be some value to the mine. She had no idea what kind of mine it had been when it had been producing. She heard about me and called one day to offer a proposal.

She wanted the mine checked out and some assays run on the ore. If there was any value to it, we would become partners in it. Well, we never could seem to get out to the mine, and she said that someone had told her it was a mica mine. I wondered about that.

A lot of time went by, and then one day we bumped into each other again. She told me she got a local Barstow man and they were running assays on the mine. I suppose I'm going to look pretty silly if they hit pay dirt.

You should visit the old mining town of Calico when you're out there. Walter Knott, the owner of Knott's Berry Farm, purchased the whole town in 1951 and began restoring it. Not a gold mining town, the boom here was for silver.

It was in 1881 that three prospectors made the discovery that was to be known as the Silver King Mine. The Silver King produced nearly $20,000,000 in its day. Borax also helped to boom Calico into life. At one time there were 2500 people living in the little town perched astride the colorful mountain.

*A view of the restored ghost town of Calico.
(Courtesy of Knott's Berry Farm)*

*Looking up the hill at Calico.
(Courtesy of Knott's Berry Farm)*

Barstow

From Los Angeles, take the San Bernardino Freeway (Interstate 10) to San Bernardino. Take the Barstow Freeway (15) to the town of Barstow. To get to Calico, continue on the freeway about seven-and-a-half miles past its junction with Highway 40. You'll see the signs; just follow them.

Oro Grande

The town of Oro Grande is outside the city of Victorville before you get to Barstow. The mines are in the hills northeast of the town. Don't waste your time panning the Mojave River, which runs through here. I have many times, and I know of many others who have, too, all with little luck. There was some surface enrichment on the hills, but it didn't seem to get washed down into the river bed.

Gold was first found here in the early 1880s, and mining was done clear up into the 1930s. The gold occurred in quartz veins and was richest near the surface. The veins were narrow and irregular. There were some rich pockets hit while mining the veins.

I spoke to an old timer here years ago (he may have passed on by now) who told a lost mine tale I had never heard before. It's not gold but turquoise that this story is about.

His father had been one of the first prospectors in the area. He found a nice vein and made quite a bit of money. He and his sons worked the mine for quite a few years until it played out.

Sometime during those early years the father had come across a rich ledge of turquoise in one of the many canyons near here. At the time there wasn't much market for turquoise, and he was a gold man, so he didn't think much of the discovery. He broke off a couple of chunks, because they were pretty, and saved them.

Years later after the gold mine had quit producing, one of the sons came across the rich turquoise ore. Turquoise had risen in price and would then have been worth mining. He went to his father to find out where the ore had come from. When his father told him he had found the ledge in one of the canyons, the son started planning a search. The father, by then an old man who

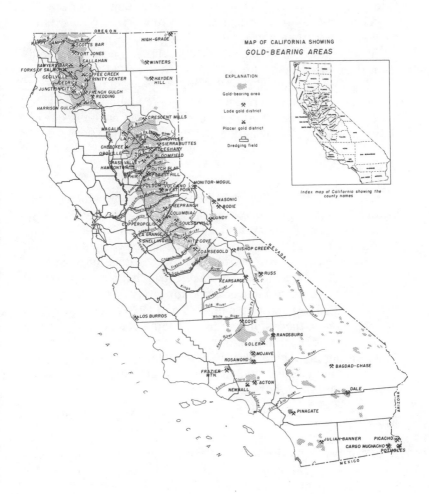

had not been out in the desert for many years, felt he could lead them right to the ledge. After several days of wandering the hills, the old man stopped and with tears streaming down his face said, "Everything's changed, . . . everything's changed."

The son took the old man back to town, where he was to pass away less than a year later.

It was the son who told me the tale. He, too, now was an old man. He showed me the ore and said that it's still there and that one day someone would find it again. I chuckled when he told me about someone who had wanted to take him up into a plane to search for the mine. "Ain't no one gonna get me up in one of those new-fangled contraptions." That was in the 1960s.

Ord

The Ord mining district is south of Barstow and east of Victorville. Entering Victorville on Interstate 15, take the Highway 18 turnoff east to Old Woman Springs Road; take Old Woman Springs Road east to Camp Rock Road. Go left on Camp Rock Road until it changes to Harrod Road. Take Harrod Road into the Ord Mountains. Harrod Road becomes a dirt road, but a good one.

The mountains are named in honor of Civil War General E. O. C. Ord. Gold was discovered here around 1870 and was worked for many years, the last major mining being done in the 1930s.

Mainly a lode mining district, there has been some successful placering done here. There are also some signs of possible large deposits of silver and copper. The gold occurs in quartz veins in granitic rocks.

Alvord

The Alvord District is east of Barstow around Alvord Mountain. The principal production has come from the Alvord mine. The strike was made in 1885 and the mine has been worked on and off since then.

Take Highway 15 east out of Barstow to Harvard Road. Go north on Harvard Road to the first good gravel road and turn right. Go right to the second road on your left. This is the Alvord Road; it will take you to the mountain.

Coolgarde

From Barstow, take the paved Fort Irwin Road to the Randsburg-Barstow Road. It is gravel here. Go left to the third good dirt road on your left. Coolgarde Camp is about three miles down the road. It is a placer mining district and has produced quite a bit of gold. The placers were first worked around the turn of the century and have been worked off and on to today. You can always get a little color here. Early dry-washing methods were so primitive that quite a bit of good gold was missed.

Grapevine

The Grapevine district is in the Paradise Mountains right above the city of Barstow. Not a large producing region, it is more famous for its rich silver mine, the Waterman, than for its gold. Both lode and placer gold are to be found here.
The gold occurs in quartz veins in the granite rocks. The veins are narrow and have not gone very deep. Only one mine has been worked very greatly and that is the Olympus. The area is to your left off the Fort Irwin Road a few miles outside of Barstow.

Goldstone

The Goldstone district is no longer open to the public as it is now a part of the United States Naval Ordnance Test Station.
There was copper as well as silver found along with gold here. Most of the activity took place from 1915 to the middle 1920s. There were several rich pockets discovered.

Historical Note

The town of Calico is a registered state historical landmark. Black Canyon, near the Opal Mountains, contains some of the most impressive Indian petroglyphs anywhere in California. The area is named Inscription Canyon and may be reached by taking Black Canyon Road from Hinkley, west of Barstow on Highway 58. There are opals, opal-agate, geodes and jasper to be found here.

*Dry placer mining in the Coolgarde District of
San Bernardino County during the early 1900s.
The picture was taken by O. A. Russell of Yermo.
(Courtesy of California Department of Mines and Geology)*

Treasure Tales

One of the most-told tales in treasure hunting concerns the Lost Gold Brick of Barstow. Sometime around the turn of the century, a gold brick was said to have been stolen from a shipment of gold bullion at the Santa Fe Railway freight station. The brick is said by some to have been worth $250,000, but most people place its value at $25,000. A gold brick worth a quarter of a million dollars in those days would have been much too heavy for one man or even two to carry off.

Years after the theft, a man reported to the Barstow police that he had been prospecting in the desert for several months and that the other man with him had confessed the deed to him. He wanted help to dig it up again and then would share the profits. The thief must have gotten wise that his partner was going to turn him in, because when the police got to the camp, the man was gone.

He told his partner that he had buried the gold bar across the Mojave River behind the freight station near a rock corral.

Oro Grande Bandit Loot

I've been told that there is some loot from a bank robbery in Needles buried alongside the Mojave River just north of the town of Oro Grande.

The robbery took place sometime in the 1930s. There were two bandits, both of whom died in prison refusing to reveal where they had hidden the money.

Calico Chinaman's Treasure

A Chinese prospector is said to have won $25,000 in gold one night in Calico. He had a claim up in Wall Street Canyon above the town. He took his winnings and went back to his cabin.

The next morning the losing gambler got drunk and mean. He told everyone that he was going to get that damn Chinaman. He went up to the cabin and killed the prospector, but couldn't find the gold. As he drew his last breath, the dying man told a friend that the gold was buried in Calico near "the big rock."

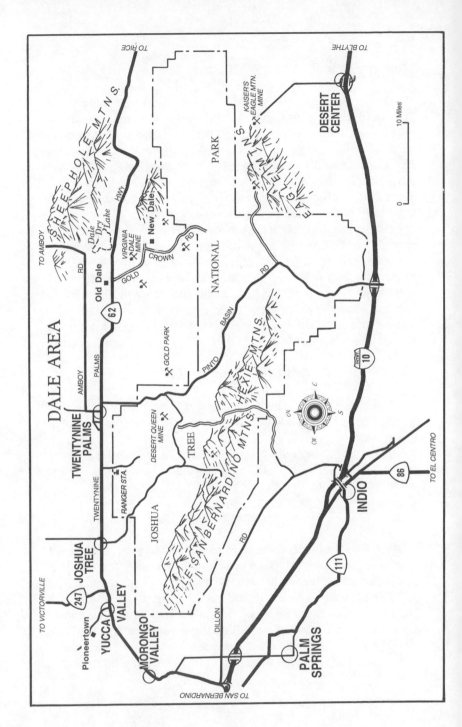

5. Dale Area

There have been three towns with the name of Dale here. The first was Virginia Dale, now known as Old Dale; next came Dale the Second, and last was Dale the Third or New Dale. All three have now vanished.

There are still things to be seen at the site of the Virginia Dale Mine, which is where Dale the Second was located, but of the town itself there is nothing standing. There are several gold bearing districts here in the high desert country. Since Dale was the best known, we will put them all in one hat and call it the Dale area.

Early mining was centered around Twenty Nine Palms because that's where the water was. The first claims were filed in the early 1870s. The Blue Jay Mine was the largest of the early mines. Its ore sometimes went as high as $100 a ton.

Placering began in the early 1880s in the hills north of Twenty Nine Palms. Miners rushed to the area and the town of Virginia Dale grew out of the desert. Not located directly in the placer fields, it was six miles from the mines. A well and an arrastra were its drawing points.

Placer gold was found in cemented gravels. Water was needed by the miners to separate the gold from the gravels.

The Virginia Dale Mine was discovered in 1885; it wasn't a single claim, but consisted of six different claims combined. Later the Supply Mine was located nearby and Dale the Third came to life.

It is well worth a trip just to see the giant cyanide tanks carved into the side of the mountain at the Virginia Dale Mine.

I have a prospector friend who wants me to investigate a find down by Desert Center. They have some ore that contains gold, silver, and turquoise in good-paying quantities, found as float in a wash near Desert Center. They haven't been able to locate the lode. I told them that as soon as this book is finished we'd see if we can track it down. If you can't find me, you'll know where I am!

Dale

From Los Angeles take any of the Freeways to San Bernardino. From San Bernardino, take I-10 east. If you are coming from the south, take Highway 60 east out of Riverside and it joins the 10 at Beaumont. Continue on the 10 until you come to the Highway 62 turnoff, past the turnoff to Palm Springs. Stay on Highway 62 until Twenty Nine Palms.

Old Dale is approximately 15 miles from Twenty Nine Palms. There is nothing but a lot of old tin cans and a couple of foundations to mark the site.

Dale the Second is south on the Gold Crown Road about four-and-a-half miles from old Dale. Look for a water tank on the right-hand side of the road. This is the Gold Crown road. It is dirt, but a good road with only a few bad ruts. The Virginia Dale Mine will be on your left as you head south.

Dale the Third is about two more miles farther down Gold Crown Road and is again on your left near the side of the mountains.

The gold occurs in quartz veins mainly in diorite and granite. Some of the veins were quite thick, as much as ten feet wide, according to reports. There were several rich pockets struck in the early days.

Twenty Nine Palms

The State of California Department of Mines and Geology puts the Lost Horse, Gold Park, Hexie, and Pinon districts with the Twenty Nine Palms district. These mines or groups of claims, as some were, are to the south of Twenty Nine Palms in the hills and mountain ranges.

Most of the mining has been lode, but you can still get some color in some of the washes by placering with a good dry washer.

The lode deposits are mostly narrow quartz veins with pyrite and a lot of iron oxide. A few of the veins contain some small rich pockets. There has been some kind of production, both large and small, here since the 1860s.

Most of the old diggings are in the Joshua Tree National Park and are now closed to mineral location.

The Gold Crown Mine in the Dale District of Riverside County. This picture, looking east, was taken about 1936 by W. B. Tucker. (Courtesy of California Department of Mines and Geology)

A fourteen-mule team.
(Courtesy of Title Insurance and Trust Company)

The other areas are still open and prospectors of a new breed with modern equipment are treading the hills once again. Maybe one day soon the cry of "EUREKA" will echo through the hills, and Dale the Fourth will be born.

Treasure Tales

Lang's Lost Loot

John Lang was a desert prospector who roamed the hills with little success around the turn of the century. He met another prospector named Diebolt in his wanderings who had made a rich strike. Diebolt wanted to sell his claim for some reason, so Lang and a partner named Ryan bought it and began mining the ore.

At one time Lang had lost a pack horse in the area of the mine, so they decided to call it the Lost Horse Mine. The Lost Horse Mine went on to become one of the largest-producing mines in the region, estimated at between one-half and a million dollars.

The mine was doing so well that they were soon running two shifts a day. Lang ran the night shift while the Ryan brothers ran the day crew. It wasn't long after the two crews started that trouble began to brew between the two partners.

When they began comparing how much gold each crew was reporting, Lang's was always several ounces less than Ryan's. Ryan soon became convinced that Lang was hi-grading him.

They had an argument that ended by Ryan buying out Lang. He then ordered Lang off the property. Lang protested and finally was led off at the end of a gun.

Lang spent the next several years searching for a new strike. He had used up all the money given him for his half of the mine and was begging drinks when the vein at the Lost Horse gave out.

Shortly after the Lost Horse closed down, old Johnny Lang began spending money like water. Where was the money coming from?

Only one man knew. Bill Keys was his name. Keys told the story after Lang died. Bill owned the Desert Queen, a working mine at the time. Lang would bring gold to Keys, who would include it in his shipment to the bank and then pay Lang. Keys figured it wasn't any of his business where the gold came from. Lang did tell him that he had lots of gold buried around his old cabin at the Lost Horse Mine. For nearly ten years, Lang would bring Keys about a thousand dollars worth of gold each time he came.

It was on one of his trips to the Lost Horse that Lang met his end. He was coming back with a load of gold when he got caught in a snow storm. He rolled up in his bag to wait out the storm. He was still laying in his bag dead when he was found nearly three months later.

They say Lang's friends buried him with his pockets full of stolen gold. The Lost Horse Mine is in the Joshua Tree National Park now.

Long's Lost Mine

L. O. Long was a prospector who had came to this area after working the rivers and creeks of the Mother Lode. He found a rich placer in the mountains north of Dale. He took out over one hundred ounces in a short time and went to San Bernardino to celebrate his good fortune and stock up enough supplies to last him for several months.

When he returned to his diggings, he fell and injured his leg. When he became too weak to work and his leg started turning green, he decided to return to San Bernardino to see a doctor there who was a friend of his. He was so sick he left everything behind him, even his shotgun.

By the time he got to San Bernardino it was too late. He died of blood poisoning a short time after he arrived. One of the last things he did was have the doctor write a letter for him to an old prospector friend. He said that the deposit was in a small brush canyon fifteen miles east of Dale Dry Lake, just below a spring.

A miner found a shotgun and other things in a small canyon in the Sheephole Mountains several years later. He didn't know about the letter until years later, and he never was able to find the right canyon again.

The Golden Volcano

There is a story they tell around the campfires here of a volcano lined with gold in the Hexie Mountains. The Indians of the region had a legend about the volcano and passed it on to the white man.

Two prospectors are said to have found it once and picked up all the gold they could carry in a matter of minutes. They went on a grand spree, telling everyone that they could get all the gold they wanted at any time. After they had spent all the gold from the first trip, they packed up their gear and headed out toward the Hexie Mountains. Maybe they talked too much, for they were never seen again.

Folks claimed that the Indians killed them because the volcano was sacred ground. The Hexie Mountains are now a part of the Joshua Tree National Park.

Points of Interest

Pioneer Town is a small town north of Yucca Valley. It was built as a western movie set and is still used once in a while for that purpose. Several television series have also been shot here.

Also near Yucca Valley is Desert Christ Park. Here you will find life-size figures and scenes from the Bible.

Little side trips are nice for your family when you're out on a prospecting trip. The little ones can become bored pretty easily climbing around the rocks all day.

Prospecting Tip

You may want to investigate the area around Morongo Valley. The area is listed as a gold district by the State Department of Mines and Geology. The mines were in the mountains west of the town. There is even a lost mine tale here, too.

The early padres are said to have had a rich mine near here which they worked for several years before being run off by the Indians.

About 1870 a miner is said to have come across the old shaft. He took samples from the vein which assayed out at $1000 a ton. Whatever happened to him or the mine, no one knows. Like so many others, he was claimed by the mountains that housed him, I guess.

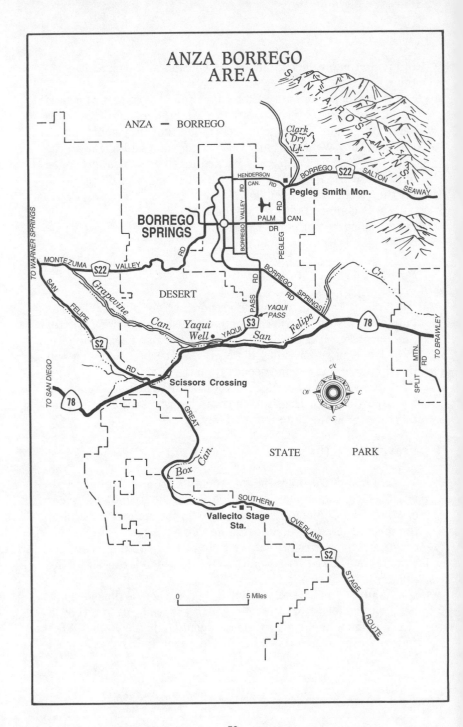

6. Anza-Borrego Area

California's Department of Mines and Geology doesn't designate the Anza-Borrego Desert as a gold-bearing area. Nevertheless, it is an important area in the story of gold in California. There is gold here, both placer and lode, and some of our most intriguing lost treasure tales are told of this region.

I learned a big lesson in desert driving here. I'll pass on my experience in the hopes that you will not make the same mistake and have to learn the hard way as I did. This all happened years ago when I had first caught gold fever.

I was returning from a business trip to Yuma and had decided to spend a little time prospecting on the way home. Figuring the best spots would be back where few people went, I scouted for dirt roads leading back into the canyons.

I was driving, of all things, a big Lincoln Continental. That poor car, I put over a hundred thousand miles on it, and took it into some of the most God-awful places, loaded down with everything from picks and shovels to sluice boxes and dry washers.

I finally spied what looked like a good road and took off down it. It was the middle of summer and the temperature was over a hundred degrees. After driving for a few miles without finding any place that looked worthwhile working, I decided to turn around and try to find another road. I found a spot that looked like people had turned around there before and started to turn around. Others might have been able to turn around there, but I tell you one thing; they weren't driving Lincolns. I got stuck!

For hours I dug sand from around the wheels, I threw everything I could find under the wheels—wood, brush, palm leaves, even my jacket—trying to get some traction. Nothing worked. Every time I would try to move the car, it would just dig itself deeper into the sand.

I had no water and was working in the heat of the day. The sand was up to the doors and I was digging with all my might, cussing my luck and getting angrier by the moment, when I must have passed out.

When I woke up, I was lying on my back next to the car and the hot sun was beating down on me. For a moment I just lay there, then I made my way to the little shade given by the car. As I sat in the shade, I wondered what I would do. Then, as improbable as it may seem, two figures came striding over the dunes.

Now get this: it was two old lady hikers. They must have been in their seventies. They were wearing desert-type pith helmets, tan army-style short pants, sturdy hiking boots, and had canteens hanging over their shoulders.

Well, these two sweet old ladies doctored me up and gave me some water, then told me they would hike to the ranger station only a few miles away to get help. A few miles...I couldn't have gone fifty feet in that heat. They gave me a little lecture on being careful in the desert and they went off over the dunes.

An hour or so later the Rangers came bouncing down the road toward me in a Ford Bronco. They hooked a cable to the front of the Bronco and to the front of my car, then backed that Bronco right up the side of the hill and pulled my big barge right out of that hole I had gotten into. Soon I was on my way home.

There you have it—a six foot three, two hundred-fifteen pound man rescued by two little old ladies. Don't take your chances in the desert; those two little old ladies might not find you. (By the way, I now have a Bronco myself.)

Anza-Borrego

From Los Angeles take the San Bernardino Freeway (Interstate 10) to the city of Indio. Take Highway 86 south from Indio to Highway 78; follow 78 west into Anza-Borrego State Park.

San Felipe Creek

Highway 78 follows San Felipe Creek part way through the park. I have seen relics uncovered by prospectors and treasure hunters in San Felipe Creek. Most of it was of early Spanish origin, such as pieces of armor.

At Scissors Crossing (Highway 78 and S-2) is the site of the San Felipe Butterfield Stage Station; here is where the Southern Emigrant Trail joined San Felipe Creek.

Warner's Hot Springs

Even though Warner's Hot Springs is not located in the desert, it is of such importance to the Anza-Borrego area that I will place its location for you. If you take S-2 north at Scissors Crossing, it will lead you to Highway 79. Go right on Highway 79 until you come to Warner's Hot Springs. Warner's Ranch is off S-2 one-half mile before it joins Highway 79.

Peg Leg Smith Monument

If you want to search for Peg Leg's lost gold, you must pay tribute to him or he will keep the golden hoard hidden forever from your eyes. It's just superstition, of course, that you must cast a stone on the mound near the monument or you will never find the gold. I admit that I did throw a rock on the pile years ago as a matter of tradition, that's all. May I lose all my lucky charms and good luck pieces if that's not true.

Take the Yaqui Pass Road (S-3) north from Highway 78 to Henderson Canyon Road (S-22). Go right on Henderson Canyon Road to the monument. Farther down S-22 you will come to the southern tip of the Santa Rosa mountains. They will be on your left as you head east.

Vallecito Stage Station

Going south on S-2 at Scissors Crossing will take you to the site of the Old Vallecito Stage Station, locale of many lost treasure tales. This is now a State of California Historical Landmark. The plaque on the monument reads: VALLECITO STAGE DEPOT (STATION). A reconstruction (1934) of Vallecito Stage Station at the edge of the Great Colorado Desert. Original was built in 1852. Important stop on first official transcontinental route, serving the San Diego-San Antonio ("Jackass") mail line (1857-1859), the Butterfield Overland Stage Line, and the southern emigrant caravans."

Superstition Hills and Mountain

The Superstition Hills are directly southwest of the junction of Highways 78 and 86. Superstition Mountain is behind the hills to the south.

Prospecting Tips

Most of the gold found here is placer. The reference to a lode deposit in the story of Peg Leg II is about the only hard rock discovery of any note. The placer gold probably came from the veins of the Julian-Banner belt and was deposited by an ancient river that flowed here millions of years ago.

There are two types of black gold to search for: the first is placer gold in black lava, the second is gold covered by a black coat. The gold is covered by black oxide from an alloy of silver or copper, with the gold that has sweated up to the surface. The lode gold of the Peg Leg II story was found in a dark-brown quartz.

I know of several prospectors who have hit rich pockets here. Look for what looks like small volcanoes; these have proven at times to contain fair amounts of gold. Try digging down in the dry washes for the pockets.

Historical Note

This area played such a major role in the early development of California that there are many historical landmarks located throughout the region. Included are Warner's Ranch, the San Felipe Valley and Stage Station, Vallecito State Station, Box Canyon, El Vado and Los Puertecitos (both are campsites of Juan Bautista de Anza's expedition of 1775), the Palm Springs Stage Station, the Butterfield Overland Mail Route, and of course Old Peg Leg's monument. There are pictographs carved by the early Indians, and fossils left by an ancient sea. Altogether, this is a remarkable area.

Treasure Tales

The Lost Ship of the Desert

This is a tale that is told over and over again. The lost ship has been placed in areas ranging from Joshua Tree National Park, to the Salton Sea, to Anza-Borrego State Park. The story begins back in the days of the Indians, when the land here was still covered by water. A ship, they say, could sail up from the Gulf of Mexico then. When the Spanish were looting Mexico, a ship already loaded with gold and pearls sailed into this region, seeking the seven cities of Cibola with its gold-paved streets and the fabulous Laguna de Oro. There are records showing that Spanish ships sailed up the Gulf, looking for an outlet to the Pacific (most early maps illustrated California as an island). Over the years, several people have seen a ship sitting among the dunes.

One person, a young man by the name of Manquerna, is said to have taken a fortune in pearls from an abandoned ship he discovered while crossing the desert. Another couple told of finding the remains of what looked like a Viking ship high on a canyon wall in the nearby mountains.

It is possible for a ship to have sailed into the region in the past, and it is possible that one could have foundered in a receding sea. Keep your eyes open, and you may be on hand when the desert decides to blow apart the sands and once again reveal that ancient monarch of the sea to the eyes of man.

The Lost Peg Leg

More have probably chased this phantom than any other lost mine in Southern California. Not really a mine but a placer deposit, its lure is as strong today as ever. Peg Leg, whose real name was Thomas Smith, was neither a miner nor a prospector. He was a man of wild spirit who roamed the west in the early 1800s. He lived with the white man and the Indian during his life time.

It was in 1829 that Peg Leg set out from Yuma for Los Angeles with another man, Maurice Le Duke by name. Where they crossed the Colorado River and what route they took has puzzled seekers of the lost gold ever since. They had only been out for three days when they realized that they had seriously underestimated their water supply. The weather was even hotter than usual and they realized that they would have to replenish their meager supply or end up as buzzard bait. To make matters worse, a severe sandstorm descended upon the two men.

Thrown off the trail by the sandstorm and becoming desperate for water, Peg Leg climbed one of three nearby buttes to look for signs of water. Tired from fighting the wind and sand and the exertion of climbing the butte, he sat down to rest. It was then that he noticed that the butte was covered with walnut-sized black rocks. He idly picked up one and was amazed by its weight. He put a couple of the rocks in his pocket, thinking he would ask someone what they were when they got back to civilization. They spent the night at the butte and the next day were able to make their way to a spring. After they had rested and refreshed themselves, they started out for Los Angeles once again.

Somewhere along the way, Peg Leg learned that the heavy black rocks were gold. For some reason Peg Leg didn't seek to return to his find at this time. Some historians think that it was because he was still a young man and too busy raising hell, and besides there was no place to spend a whole lot of money then. At the time, Los Angeles was a sleepy little pueblo of about a thousand souls.

It was twenty years later when the rush to the Mother Lode began, that Peg Leg realized the value of his discovery. He returned to the desert to reclaim the fortune, but it was too late. Time and nature had altered the landscape and his memory. He never could locate the black butte with the chalky yellow base again. Peg Leg used Warner's Hot Springs as a base for his search, which he centered in Borrego Valley. After a few years, he gave up and went north to live with the Indians.

He is said to have concentrated his efforts on horse-stealing and gambling while boasting of the fortune he had in his hands for a moment that one fateful day. He died in San Francisco in 1866 at the age of 65. Since his death, the legend of Peg Leg's gold has continued to grow. No one knows how many men have risked everything in hopes of finding this lost bonanza.

One man may have found it. For the past several years, someone had written to *Desert Magazine* claiming to have found

The Fricot Nugget. This 201-ounce (troy) cluster of gold crystals is on display in the Division of Mines and Geology mineral exhibit in San Francisco's Ferry Building. Melted down as gold, it would be worth several thousand dollars, though its value as a historical object and museum piece is much more. The nugget is shown here as slightly less than half its actual size. Photo by Mary Hill. (Courtesy of California Division of Mines and Geology)

the Lost Peg Leg Mine. He backed his claim by sending one of the black nuggets with the letter. I think that he was telling the truth. Nevertheless, when you are out that way, stop and doff your hat to the monument...to a legend.

Peg Leg, The Second

The tale of Peg Leg II begins shortly after the death of the original Peg Leg Smith. This Peg leg was said to have lost a leg in the Civil War and since his name was Smith, too, he became known as Peg Leg Smith, the Second. This Peg Leg was a much different man than the first. He never lost his gold. He discovered a rich ledge of chocolate-colored quartz studded with gold. He was well-known in Yuma and San Bernardino, where he was a free spender.

Many times he would be followed from those towns by men who were eager to share his strike with him. He would wander through the hottest parts of the desert when he knew he was being watched. While other men would crack under the blazing sun, their lips parched by thirst, he seemed to thrive in the heat. He worked his mine for years, coming into town every so often for a spree.

Then one day a man who was said to be a deserter from the Army post at Yuma found Peg Leg's body, lying dead along the trail. A fifty-pound bag of rich gold ore was beside him.

The soldier, who was lost himself, somehow made his way to San Bernardino. Dehydrated, delirious, cut and bruised, he was hospitalized, but still brought several pounds of ore with him. The ore was identified as Peg Leg's, but the soldier never recovered and died in the hospital. Before he died, he revealed what he knew about his discovery to a doctor, who made several trips into the desert using the clues provided by the dying soldier, but was never able to find the mine.

Chances are that Peg Leg hid the outcrop and all traces of his labors each time he left them. The rich brown quartz is still there. Maybe time and the elements will someday reveal this treasure to you.

Montezuma's Treasure

If you find this one, you'll be as rich as the Count of Monte

Cristo. This story begins in Mexico, when Spanish soldiers under Cortez were plundering the Aztec Empire. Mexican legend says that Montezuma, to foil the Spanish, ordered the bulk of the Aztec gold, silver and gems be taken north farther than any man had ever traveled and hidden there until the Spanish were vanquished and driven from Mexico.

The Indians of the desert told of the great army of Aztec warriors and slaves who came into this area and were lost in the area of Superstition Mountain. This was when the low desert was still under water and Superstition Mountain was an island in the Gulf of California.

Geologists claim that it is very possible that there is a large limestone cave honeycombing Superstition Mountain. I have a treasure hunter friend who feels he knows an entrance to the cave and wants me to accompany him on a search for this great treasure. That's a trip I am going to take.

Vallecito Treasures

Shortly after the stage station was abandoned by the Butterfield people, a Mexican bandido and his wife and servant girl took over the station building as a hideout. For several months, he victimized the miners returning to the East with gold they had wrested from the Mother Lode. Soon he had eighty thousand dollars in stolen gold.

Somehow he learned that the law was about to raid his hideout. He sent his wife on ahead of him to Mexico. He kept the Indian girl with him to do the cooking and to gather their things together. The bandit took the gold and loaded it onto a pack horse, telling the girl that he would return shortly.

Hours later, the man's horse returned to the hideout, but no sign of the bandit could be seen. The girl backtracked the trail of the horse and found her master lying dead by the side of the trail. His horse had thrown him and he struck his head on a tree.

When her husband failed to arrive in Mexico, the wife returned to the stage station. She found the Indian girl still there. After she learned what had happened, she set out with the girl to locate the gold he had taken with him. They found the pack horse in a small canyon near where the bandit had died, but its packs were empty and a shovel had been tied to the packs. They searched for several days, but could never locate the spot where the bandit had apparently buried the loot before meeting his death. This area is now called Treasure Canyon.

An Indian known as Sonora Joe was said to have worked a rich placer in the Vallecito Mountains. He would come into Julian with some good-sized nuggets to spend. He said that there was a cave nearby that contained many relics left by the early Indians. He just vanished one day on his way to Julian.

Bluebeard's Treasure

Bluebeard Watson was a bigamist who confessed to killing fourteen of his twenty-five wives. He stole at least $200,000 from the estates of his various wives and buried much of it somewhere near Borrego Springs.

Warner's Ranch Treasures

There are several stories of lost gold centered around Warner's Ranch. The first is the tale of the wife of one of the men who worked at the Ranch. An Indian squaw who worked for this lady offered to show her a place where there was a great deal of gold. One day they set out to find the gold, but were stopped by the squaw's husband. The squaw did manage to tell her the place was in the mountains near Oak Grove.

A Yaqui Indian who worked at the ranch and always had lots of gold to spend is the hero of another tale. Whenever he seemed to need any money, he would set out down San Felipe Creek and be gone for three days, returning with a bag full of gold nuggets. When he died, nearly $4,000 in gold (encased in black lava) was found under his bunk.

These are just a few of the tales told of lost gold and buried treasure around the campfires in Anza-Borrego State Park. Try your metal detectors in Carrizo Wash, Yaqui Pass, San Felipe Creek, Grapevine Canyon, or around Yaqui Well and the Vallecito Stage Station.

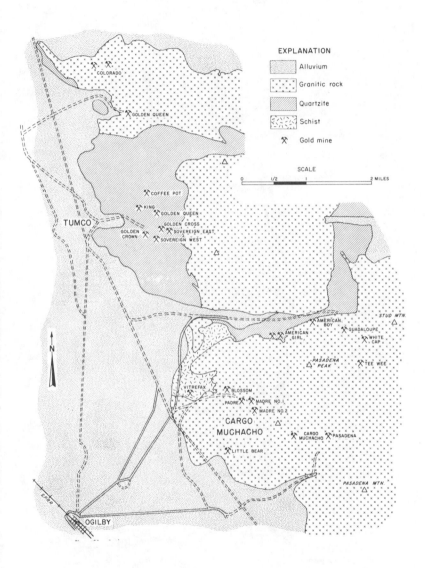

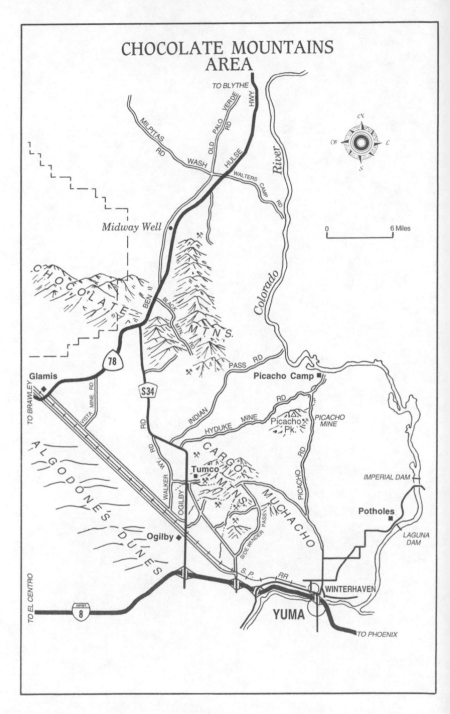

7. Chocolate Mountains - Tumco - Potholes Area

For the past several years, this area has been the most productive gold region in southern California. I met an old timer in Julian a couple of years ago, who told me that there had been quite a bit of new gold washed down out of the Chocolate Mountains. I rushed out there as soon as I could, and, by golly, he was right. There are lots of claims staked out now, but there are still many good spots left. The area is so remote, the terrain so rugged and the climate so harsh that it has not been over-worked as so many other areas have.

For a little relaxation at night, you can watch the dune buggies racing over the sand hills down by the canal. One has to wonder what the old time burro prospector and others would think if they could see how easily these machines have conquered the sands that trapped so many of their fellow prospectors in the past. Time changes things.

The Chocolate Mountains

From Los Angeles, take the San Bernardino Freeway (10) all the way to Indio. Take Highway 86 south to Brawley. From Brawley, take Highway 78. The mountains on the left after you pass the sand hills are the Chocolate Mountains. The area of the Chocolate Mountains for about 10 miles past Glamis is closed. This is for your safety, as the area is used as an aerial gunnery range. The Mesquite gold mining district is included in the Chocolate Mountains.

Lode and placer mining has gone on here since the 1800s. There were several attempts to mine the placers by large scale dry washing in the 1930s, but none were successful.

I have done the best by working the washes between Glamis and Midway Well with my electrostatic concentrator.

Golden Cross Mine, Cargo Muchacho District. This view of the mine, at Tumco, Imperial County, looks west. The photo was taken in about 1915. (Courtesy California Department of Mines and Geology)

The lode veins are quartz, containing gold and silver with iron oxides occurring in granite rocks. The veins are not large, but have produced some rich pockets.

Cargo Muchacho-Tumco

From Glamis take the road south to Ogilby. The mountains on the left as you look south are the Cargo Muchacho mountains. The mines are back up in the mountains. The Tumco mines are about two miles north of the Cargo Muchacho mines. There are good dirt roads leading to each of the areas. This is the oldest mining district in California, having been worked as early as 1780 by the Spanish. Large scale mining was done here almost continuously until 1941, when all the mines were closed.

Most of the gold mined here has been hard-rock, but you can find some placer gold in the washes. The veins here have been wide and high grade. The gold occurs with silver and sometimes copper in quartz with calcite. Try your dry washer in Jackson Gulch and you will get some color.

Potholes and Picacho

The Potholes and Picacho districts can be reached out of the town of Winterhaven on this side of the Colorado River from Yuma. Some historians state that the Potholes region is the location of the first gold mining in California. A placer gold region, the gold is found in depressions or pot-holes, thus the name. It was mined mainly by the Indians and Mexicans. There was a great deal of activity here during the latter part of the 1800s when there were up to a thousand men working the deposits. Potholes is located just west of the Laguna Dam.

The Picacho road runs north out of Winterhaven 18 miles to the old camp of Picacho. The camp is set down in the Picacho Wash. The Picacho Mine, which was a fairly good producer, was back up in the mountains near Picacho Peak. The placers were mined as early as 1780, according to records. Most of the washes here have gold. You can still see the tailings from the early Mexican and Indians efforts to dry wash the deposits. In the 1890s there was an unsuccessful attempt to hydraulic mine the gravels.

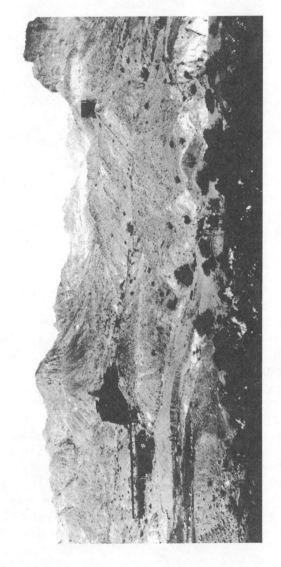

Picacho Mine, Picacho District. Picacho Peak rises in the right background in this 1921 view of the mine, in Imperial County. Photo by Ralph Baverstock, from collection of Dr. Horace Parker. (Courtesy California Department of Mines and Geology)

The gold in the placer deposits is very fine and seems to occur in small concentrations erratically placed throughout the dry washes. The lode deposits are low grade but large. The last ore shoot worked in the Picacho mine was said to be 150 feet wide and 250 feet long.

Treasure Tales

The Lost Yuma Indian Gold

There are several versions to this tale, with a varied cast of characters, but the main points are the same. Sometime back in the 1800s, a resident of the town of Yuma took in a young member of the Yuma Indians and treated him with great kindness. As the months passed, the Indian boy learned that white men attributed great value to the yellow metal called gold. The boy decided to show his gratitude by telling this treasure tale.

When he was a little boy, he and his father were following an old Indian trail between Potholes and Picacho after a violent rainstorm. His father left the trail and went up a wash. The bed of the wash had contained a wide vein of ore laced with fine wire gold. The storm had uncovered the vein, and his father said that they must cover it again. The man never touched the gold and placed rocks over it until no trace could be seen. He told his son that this was the source of the gold of the Yumas and that it was each member's duty to protect it so that no white prospector could ever find it.

The young Indian took his benefactor up the Colorado river to a place known as Ferguson's Flat. From there they walked for about three or four miles. They were walking along when the boy leading the way collapsed. The young Indian had not been well, and the excitement must have been too much for him. They made their way back to Yuma, and the boy was placed in the hospital. About a week later the boy vanished from the hospital.

Did the other members of the tribe claim him just to keep him from revealing the location of the gold, or did he just run away fearing the wrath of the tribe? The man returned to the spot where the Indian had collapsed and tried to locate the vein, but never could.

Lost Dutchman of Picacho

This area has a lost Dutchman Mine of its own. Not to be confused with the Lost Dutchman Mine of the Superstition Mountains in Arizona, its location is in the region between Potholes and Picacho.

Around 1910, an old Dutch prospector found a rich ledge of gold in a basin near a water hole in this region. Right after he had made the discovery, his mules got into his supplies, and he was forced to leave his mine and head for Yuma to replenish them. He gave the ferryman who took him across the river threepieces of ore that contained $62 worth of gold. They figured that the ore would assay out at about $20,000 a ton. The old Dutchman told the ferryman that he was going to Quartzsite first to pay off the man who had grubstaked him. He was never seen again. Old timers figure the mine was somewhere near Rochester Basin.

The Golden Sand Dunes

The sand hills fronting the mountains on the west are known as the Algodones Dunes. They are almost six miles wide in places and nearly forty miles long. They are so much like the Sahara Desert that they have been used to portray that barren waste many times by movie companies from Hollywood.

Our tale of lost gold begins in 1917 when a prospector, whose name has been lost in time, set out from Mexicali heading for the gold fields in the Cargo Muchacho Mountains. He had to pass through the dunes to get to the mountains. While crossing the sand hills, a violent wind storm came up and caused him to be separated from his burro. The burro had all his supplies including his water. He had no choice but to head to the nearest place of safety, the town of Ogilby. It took him several days to reach Ogilby; it was too long, and the desert had beaten him.

He died shortly after being found unconscious by the railroad tracks outside the town. The prospector had 17 pounds of gold nuggets on him when he was found. He told his rescuers that he had found the nuggets in a place where the wind had blown away the sand revealing the bedrock.

*A prospector with his burro.
(Photo courtesy of Title Insurance and Trust Company)*

No one has ever found this rich deposit since. The sand hills are ever changing here, as the wind creates new hills each hour, while carrying away the ones of yesterday.

Lost Gold of Mission San Pedro y San Pablo de Bicuner

This may be the oldest lost treasure in California. Around 1780 the Spanish established the mission San Pedro y San Pablo on the rich gold placers of the Potholes district. If they knew the gold was there and built the mission there for that reason, or if they discovered the deposits later, no one knows. What we do know is that the Spanish forced the Yuma Indians to mine the gold for them. The Indians suffered under this golden yoke for several years, until one day they rose up in revolt and slew their Spanish masters. They destroyed all evidence of the mission and hid the mines. The gold was either buried nearby or thrown in the Colorado River, according to the legends.

A humorous sidelight to this story took place in the 1830s when officials from the pueblo of San Diego financed a search for the gold. When the searchers reported back to San Diego empty-handed, they were promptly thrown in jail and accused of stealing the gold. Remains of the old mission are almost impossible to locate today.

Black Butte's Gold

There are many tales of lost gold throughout this whole region that are associated with a black butte. Some place the Lost Peg Leg Mine here.

One such story deals with an Indian woman who staggered into a railroad crew's camp near Glamis one day. The woman was suffering from exposure and died soon after reaching the camp. She had a bag full of gold nuggets which she said that she had found near a black butte. I would say that if you should come across any black butte around here, you would be wise to check it out very carefully.

8. Other California Desert Gold Bearing Areas

In the preceding chapters, I have covered the principal gold bearing districts of the desert with the thought of convenience to those living in the metropolitan areas. Most of the mountains fronting the California desert on the east as you near the border of Arizona and Nevada contain gold in various amounts. In most cases the roads are poor and services are not convenient in these areas.

Whipple

The Whipple Mountains are in southeastern San Bernardino County. Gold and copper in quartz veins have been mined here with some success.

New York

Also known as the Vanderbilt District, this district is in northeastern San Bernardino County, in the New York Mountains. Gold was first found here in 1861 and mining was done intermittently up to 1941. The district was important enough that a spur was built by the railroad to serve it.

The Vanderbilt mine was the principal mine of the district, with a shaft that reached a depth of 400 feet. The gold occurs in quartz veins that sometimes also contain silver and copper.

The Castle Mountain or Hart District is nearby. It had a brief boom around 1910 and was served by the same branch of the railroad as the New York District.

Ivanpah

This district covers the Ivanpah and Mescal Range of mountains. The Mollusk Mine used to be the principal producer, with

*The oak of The Golden Dream.
Site of first recorded gold discovery in California.*

a production of over a quarter-million dollars. The gold is found in quartz and mineralized breccia. The Mollusk vein was odd in that it was dolomite.

Clark

The Clark Mountains are north of the Ivanpahs and have been a source of gold since the early 1860s. The Mountain Pass Mine is currently a large producer of rare earth minerals. Gold there is found in quartz, barite and mineralized breccia.

Halloran Springs

Halloran Springs is 12 1/2 miles east of Baker off Highway 15. The Telegraph Mine, which was not discovered until 1930, is about two miles east of Halloran Springs. The gold is found in quartz veins, some as wide as eight feet, and in spots has been very rich. Indians have mined the turquoise deposits here for hundreds of years.

The Shadow Mountain District is just to the north. There have been several gold quartz veins (with a little return) worked in the Shadow Mountains over the years.

Old Dad

Just to the south of Halloran Springs lie the Old Dad Mountains. First worked in the 1890s, they had their greatest period of activity in the thirties and early forties. There has been some high-grade ore found here and the veins have been good size, from one to six feet in some cases.

Providence-Hackberry Mountain

Both the Providence Mountains and Hackberry Mountain have had small booms. The Providence (or Trojan) District enjoyed its greatest period of activity from about 1912 to 1919. The Providence Mountains can be reached by taking the Essex Road out of the town of Essex. An interesting side trip here can be made to the Mitchell Caverns, which contain some of the most spectacular displays of cave coral, stalactites and stalagmites seen anywhere in the world.

To reach Hackberry Mountain, take the Lanfair Road north from the town of Goffs. Goffs is east of Essex before you come to Needles. The gold deposits here have been mined since 1890. Recently there has been some excitement in these parts over cupro-descloizite, a rare vanadium-bearing mineral. Vanadium is used as an alloy to toughen steel, and up to now our principal source of it has been Peru. The Arrowhead District is also in the Providence mountains, and the early Mexican miners did very well working the rich surface ores here in arrastras. The Hidden Hill Mine in this district once hit a pocket that weighed 300 pounds and was worth $13,000.

Clipper Mountains

Known as the Gold Reef District, gold was first discovered here around 1915. The mining area can be reached by taking the Danby Road north from old Highway 66 about six miles west of Essex.

There is a lost treasure tale told of the Clipper Mountains. The story begins in 1894 when a Santa Fe railroad employee named Schofield stumbled across a Dutch oven full of gold while searching for underground water. He had taken a day off from his labors and was just looking around when he made his discovery.

He found an old trail and decided to follow it to see where it led. It took him through a gulch, past a spring bubbling from the side of the gulch, over three low hills, and into another canyon. It was somewhere near here that he saw an opening in the cliffs.

He made his way through the narrow opening into a mass of black rocks. Here was an ancient mining camp. The mine shaft was nearby and still in good condition. While wandering around the old camp, he tipped over a beat-up Dutch oven left there. His eyes bulged with excitement when the cover fell off and gold nuggets spilled out on the dirt at his feet. Schofield filled his pockets with all the gold he could carry and made his way to Los Angeles to celebrate. When he tried to return to the camp, he was never able to find it again.

Some treasure hunters believe that the mine was not in the

Adobe ruins.
(Courtesy Title Insurance and Trust Company)

Clipper Mountains, but in the Old Woman Mountains to the south. The Old Woman Mountains are another gold district with a small amount of production.

Arica-Riverside-Whipple Mountains

All three of these mountain ranges have produced a fair amount of gold over the years. The Riverside Mountain District is also known as the Bendigo District. Some of the deposits there were quite large, with veins 15 feet wide. While in this region, you may want to spend some time searching for California's second most sought-after mine, the Lost Arch. The Lost Peg Leg is first.

Earle Stanley Gardiner, author of the Perry Mason detective books, employed airplanes and helicopters in his search for this lost treasure, without success, I might add.

There are two versions to the tale. The first tells of a party of Mexican miners on their way to work the placers at La Paz, Arizona. They camped one night in the Turtle Mountains and discovered a rich placer field. They agreed to stay and work these diggings rather than go to La Paz. A couple of stucco cabins were built to house the men. The cabins had an arch connecting them and made a sort of third room. They worked the deposits until the water ran out, a period of three to four months.

They accumulated $30,000 to $40,000 in gold, splitting the gold into equal shares, returning to their homes and making a vow never to return without each other or to reveal the location of the mine to any outsider. They never did return.

The other version tells of two prospectors named Fish and Crocker who ran out of water and were seeking a fresh supply. They had separated and Fish went up a canyon past some giant boulders into a wash. Over the wash there was a natural bridge or arch. Fish sat down in the shade to rest for a moment. After a while, he began to dig in the sand looking for moisture. He didn't find any water, but he did find gold, lots of gold. He gathered as much as he could carry and returned to camp to meet his partner. He showed Crocker the gold and they agreed to head for the Colorado River to replenish their water and then return and work the placer.

They did reach the river; when they did, Crocker consumed

too much water too soon and became violently ill. Fish tried to get him to Ehrenberg, on the Arizona side, where there was medical help, but Crocker never made it. After Crocker's death, it was several months before Fish felt like returning to his Arch Mine. When he did try to return, he could never find the arch again. Fish spent the next seventeen years searching for the gold. He died in 1900 in San Bernardino, a poor man.

The Lost Arch has been seen by many people over the years, but it is always later that they learn what they might have had. There is one other clue to this treasure: red hematite. A large outcropping of hematite is said to mark the spot also. Hematite is an iron ore and is sometimes a good sign for gold.

Orocopia Mountains

The Dos Palmas District in the Orocopia Mountains has been the source of small amounts of gold in the past. Several narrow quartz veins containing some gold were worked in the 1890s. The Orocopia Mountains are northwest of the Chocolate Mountains.

Chuckwalla

The Chuckwalla Mountains are northeast of the Chocolate Mountains. Gold was first found here in the 1880s and mines were worked into the early 1940s. The quartz veins have been very rich in places and sometimes contain silver, copper and lead along with the gold.

Somewhere along a trail leading through these mountains, a prospector hid two bags of gold nuggets. They were the same type of black nuggets found by Peg Leg Smith. If you come across a couple of bags of black rocks, don't pass them by. They could be worth ten to fifteen thousand dollars.

Mule Mountains

No work has been done here for quite a few years. The mountains are to the southwest of Blythe. Gold and copper in quartz veins in granitic rocks were the source of the production.

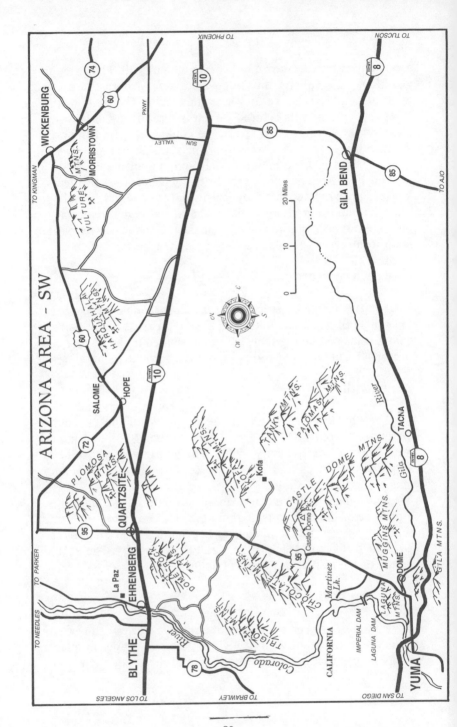

Arizona Areas

Directly across the Colorado River from the Potholes district are the Laguna placer deposits of Arizona. Even though the gold was probably deposited at the same time, the California Potholes district is much more extensive. Potholes located near what is now the Laguna Dam have been found to be the richest source of placer gold on the Arizona side as well. Some of the potholes were up to 100 feet above the river. There were several large nuggets found by workmen building the dam. A fair size gold-quartz vein was also discovered at the time.

An effort to dredge the deposits was made in 1884. It went down to defeat the next year when the dredge was sunk during a flood.

The gold found here is very coarse and rust-colored. Geologists feel that the gold is derived from irregular quartz veins in the masses of black schist and granitic gneiss that rise high above the river.

There are several other good placer locations around the Laguna Mountains. The McPhaul placers are ancient Tertiary gravels found near the southern tip of the range close to the Gila River and about a mile north of McPhaul Bridge. Las Flores Gulch in the southeastern part of the range (a little over a mile north of the Gila River) has produced a small amount of gold in the past.

Gila City Placers

The Gila City Placers are located twenty miles east of Yuma at the northern end of the Gila Mountains. The placers were first discovered around 1856. By 1858 there were miners all over the region working the gravels. The most productive area has been Monitor Gulch. Monitor Gulch gets its name from the efforts of a San Francisco company to hydraulic mine the placers in 1870. Another attempt at mining the gravels was made in 1931 by pumping water from a well near the railway. Both op-

erations were short-lived and mostly unsuccessful.
Some large nuggets have been found in benches in the gulches west of Dome. Practically all the gulches from an area 1/4 of a mile east to three miles west of Dome contain gold. The deposits are spotty with some high-grade pockets.

The Muggins Placers

The Muggins Mountains contain several gold placers. Burro Canyon has been the most productive area. There have been several high-grade pockets in some of the other canyons.

Each rainy season seems to produce new gold in the washes. The gold is washed down from the conglomerates on the canyon walls. Not worked as much as some of the other placer deposits of the region, they offer the prospector a greater opportunity for success than most areas.

Castle Dome Placers

The Castle Dome Mountains are 31 miles northeast of Dome. There are placer deposits in several of the gulches and canyons in this range, but you have to dig pretty deep to find them. The gold is mostly all near or on bedrock in the washes. They were first worked around 1884.

La Paz Placers

The La Paz placer deposits are at the western foot of the Dome Rock Mountains, about six miles east of the Colorado River. Production has been hampered by a lack of water, as is true of almost all the placer deposits in Arizona. One of the richest fields in the state, a great deal of gold and some fairly large nuggets have come from here. The Indians knew of the gold here for many years, but it was not until 1862 that they revealed the location to the white man.

Pauline Weaver, the famous scout and trapper, and some others were shown the deposits by a friendly Indian. They took out over $8,000 in a few days. When they went to Yuma to get supplies, word of their discovery leaked out. The rush was on.

Miners came from all over to work the rich gravels. The town of La Paz sprang up about two-and-a-half miles from the river. It soon boasted of a population of over 1500. La Paz became a regular stop of the Overland Stage and served as county seat until 1871. Today, nothing remains of the town but a few crumbling adobe walls.

Over a million dollars in gold is estimated to have been taken out the first year, and that was when gold was selling for only $15-$17 an ounce. One nugget found by a man named Juan Ferra weighed 47 1/2 ounces. Water was so scarce during the rush that it sold for five dollars a gallon.

The gold is found in gravel deposits in Goodman Arroyo and Arroyo La Paz, and in the tributary gulches running into the Arroyos. Ferrar Gulch has produced the most gold.

Trigo Placers

Farther south on the western slope of the Dome Rock Mountains are the Trigo Placers. Several attempts to dry wash the firmly cemented gravels over the years have met with little success.

Plomosa Placers

The Plomosa Placers are in the washes and arroyos that empty into the La Posa Plain between the Plomosa Mountains and the Dome Rock Mountains. The most productive placers are the La Cholla, Oro Fino, Middle Camp and Plomosa. The La Cholla, Oro Fino and Middle Camp Placers are near the Dome Rock Mountains. The Plomosa Placer deposits lie near the Plomosa Mountains.

There have been several large scale operations attempted here with only moderate success. The gold is found mainly on or near bedrock, but in some places it is scattered all through gravels.

Kofa Placers

The only known placer deposits in the Kofa Mountains are in a gulch below the King of Arizona Mine. The placer gold found in the gravels is not the same as the gold in the lode veins of the King of Arizona or the North Star mines. It may come from some low grade veins up in the mountains or maybe there is a rich vein still waiting to be discovered here.

You have to dig deep here or find a spot where bedrock is exposed because that's where the gold is. The gold is very coarse and there are some fair size nuggets.

Tank Mountain Placers

There are several gulches on the eastern slope and in the northwestern portion of the Tank Mountains that contain placer deposits. Try your dry washer here for some color.

Harquahala Placers

Some gold has been taken out of placers in Harquahala Gulch in the Harquahala Mountains. This was early in the 1880s. This has been a rich hard-rock mining area.

Vulture Placers

One of the best known mines in Arizona is the Vulture Mine. Discovered by Henry Wickenburg in 1863, it produced over 300,000 ounces of gold. It was said that Jacob Waltz (who has gone down in treasure-hunting lore as the Dutchman of the Lost Dutchman Mine stories) worked here, but was fired for high-grading.

To reach this location, go west from the town of Wickenburg on Highway 60 about eight miles to the gravel road marked "Vulture Mine." Follow this road another fifteen miles to the site. The most productive area is in Vulture Wash; the deposits are in the gravels beginning at Red Top Basin and stretch about two miles. This area has produced some large nuggets and is a good area to metal detect as well as dry wash.

San Domingo Wash

This is a popular area for dry washing, and is a nice one-day outing from Phoenix. San Domingo Wash, an eastern tributary of Hassayampa Creek, runs out of the Wickenburg

Mountains north of the town of Morristown. The most active period here was in the 1870s. Old Woman Gulch was a large producer at that time.

To reach this location, take Highway 60 northwest out of Phoenix about thirty-two miles to Morristown. There is a good dirt road going north from Morristown that will take you to the deposits. This is still a popular area for dry washing.

Two miners take a break around 1910 at Esmeralda Camp, north of Wickenburg.

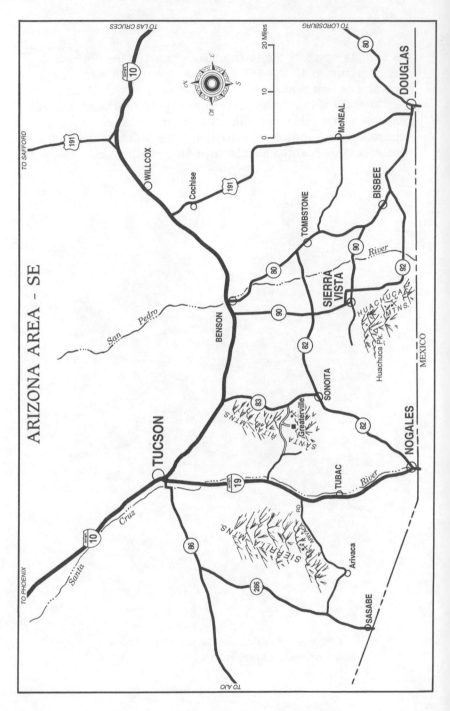

Las Guijas Placers

This district is located southwest of Tucson, near the town of Arivaca. To reach this area, go south out of Tucson on Highway 19 about thirty miles to Arivaca Road, then turn west. The deposits here have been worked off-and-on since the 1860s. Most of the washes contain some gold. Most of the gold is very fine, but some nice nuggets have been recovered here.

Greaterville Placers

This area is located on the eastern foot of the Santa Rita Mountains and has been one of the largest producing regions in Arizona. Take Highway 10 east from Tucson about 22 miles to Highway 83; go south on this road to the Greaterville Road. Turn west here and continue to Melendrez Pass Road. Follow Melendrez Pass Road south to Greaterville. This will put you right in the middle of the gold bearing areas. It's a good place to nugget shoot, as quite a few nuggets have been recovered here. The largest nugget found here weighed 37 ounces.

According to reports, it was common in the early days to get an ounce of gold a day. All the gulches in this area contain gold. Placer deposits are located mostly on the lower east slope of the Santa Rita Mountains. The richest gravels are those along stream courses, although placers are also found in older gravels on benches and even tops of ridges. It's worth a trip for the history and the pretty country—as well as the gold.

Ash Canyon

This district is located in the southeastern portion of the Huachuca Mountains. To get here, go south on Highway 90 from Benson to Sierra Vista, then take Highway 92 to Ash Canyon Road. This is a good dirt road and will take you to the placer deposits in the canyon. Some large nuggets have been found here. You can dry wash and nugget shoot here.

Prospecting Tips

None of these districts has any water to speak of except when it rains. You need a good sturdy dry washer to work the gravels as they are very angular and coarse.

Most prospectors only work the deposits during the cooler seasons of the year. Trying to work the deposits in the summer will give you a real good idea of what Hell must be like.

If you have a little rockhound in you (and most prospectors do), the Castle Dome mine dumps are famous for their collecting material. You can find calcite, cerrusite, and vanadinite here.

The area around Quartzsite contains Apache tears, agates, petrified wood, onyx, obsidian, and some beautiful quartz crystals. Quartzsite is also the home of the Quartzsite Rock Powwow held every year.

The shaft of the Kelvin Sultana Mine in Pinal County.
(A. L. Flagg Collection)

Treasure Tales

The Yuma Indians of the Colorado River region have a legend of a rich lost placer in the Castle Dome Mountains with a curse on it. The story begins in the days of the first Spanish visitors to the area.

A band of young braves set out from the main camp to hunt mountain sheep in the mountains. After searching for several days without success, they found that they had wandered into an area that none of them had ever seen before. Their water supply was nearly gone when they decided to give up the hunt. Some of the braves were selected to search for a water hole in the many canyons running out of the range.

One of the young men returned and told of finding a canyon whose entrance could not be seen except from a certain place. Inside this strange canyon, the walls were an odd color such as he had never seen before. There was a small stream and a pond. Not only was there water but he said that there were tracks showing that many animals came here to drink.

As they entered the canyon, several of the braves became frightened by the orange glow of the canyon walls as they were struck by the rays of the setting sun. Their thirst and the possibility of a good kill caused them to put aside their fears. After they had refreshed themselves, the Indians made blinds to conceal themselves from the game.

During the night they were able to bring down several deer and sheep. By morning they had all the game they could carry. As they were cleaning up in the pond after dressing the meat, they noticed that one edge of the pool nearest the canyon wall had crevices filled with gold. The Indians gathered as much as they could, knowing the Spanish were eager to trade for the yellow metal.

The sky was beginning to darken rapidly. Knowing how fast a desert storm can break, the Indians hurried to get out of the long narrow canyon. It was too late. Before they had gone but a little way a deluge of water fell from the sky. The cloudburst created a flash flood that swept over the fleeing Indians. Only one of the hunting party escaped with his life.

The battered and bruised brave made his way back to the

village and told the elders what had happened. Somehow he had managed to hang on to a small bag of gold nuggets to prove his story. The medicine man told the tribe that this was an evil place and that forever it would be haunted by the souls of the braves who had died there. From that day on the canyon was a forbidden place to the Yumas.

Over the years its location has become lost. If a modern day Yuma Apache did not believe the old taboos, he still would have to hunt for it, just like you and I would.

Nummel's Lost Mines

John Nummel was a prospector who roamed over the mountains in this region around the turn of the century. He told of two strikes he'd made and then lost.

The first was a ledge of dirty yellow quartz laced with fine gold. He was walking somewhere near the Yuma Wash between the Trigo Mountains and the Chocolate Mountains when he sat down under a tree to rest. He noticed a ledge of yellow quartz next to him and, being a prospector, he took a sample. The outcrop was rich. He planned to come back and work the ledge, but he could never find it again.

The second tale is not of gold but of silver. Nummel had been on one of his prospecting trips into the Trigo Mountains and had picked up a bag full of samples to check later. He forgot about the bag of samples (God help us, we've all done this), and sometime later he just dumped the ore in his yard. Another prospector friend of his was looking through them one day and got very excited. "John, this is some of the richest silver ore I've ever seen," he exclaimed. "Where did you get it?" "Let me think," John Nummel murmured. Think as he would, John Nummel could never remember where in the Trigo Mountains he had picked up the rich ore.

There are several tales relating to a lost silver ledge in the Trigos. One extremely rich lode is said to be somewhere north of Clip Mountain. It has been found four or five times over the years, but for some reason or another no one has ever been able to go beyond that. Maybe it's one of those that a curse has been put on. Who knows?

Lost Bicuner Gold

There are those who claim that the lost gold from Mission San Pedro y San Pablo de Bicuner is buried on the Arizona side of the Colorado River. (Most stories place it near the mission site in California.) According to them the gold was taken across the river by the Indians and buried near Squaw Peak (Sugarloaf).

Lost Gold of the Gilas

Somewhere in the Gila Mountains is said to be a rich gold mine first worked by the Spanish and then by a group of Frenchmen. The Frenchmen had been on their way to the gold fields of California when they discovered a rich Spanish mine. The Frenchmen guessed that the Spanish miners had been killed or run off by the Indians. They worked the mine for several months. When they had collected enough gold to make them rich for life, they left the mine. Before they were even clear of the range, Apaches attacked the party. Caught by surprise, the small band of miners fell quickly.

That day and repeated many times throughout history, gold caused the desert sands to be stained red by the blood of men.

Years later, three prospectors found the skeletons of the Frenchmen and gold all around them. They were never able to locate the mine. Of the three miners who found the massacre site, only one survived to tell of it. One of the men went crazy and killed one of the others. He then was shot by the third man.

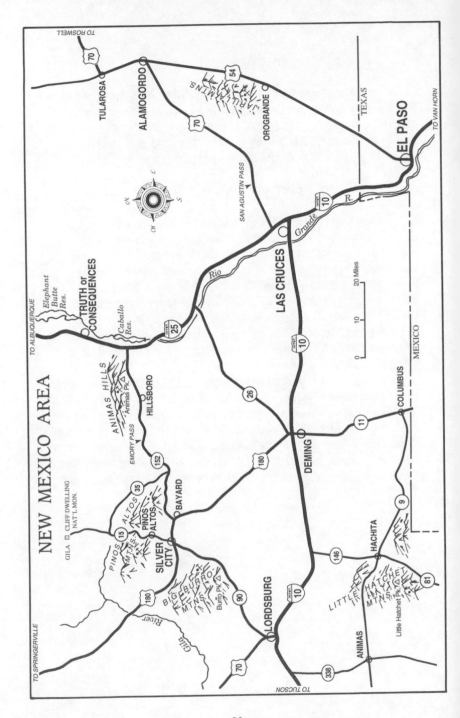

New Mexico Areas

The state of New Mexico has had the seventh-largest gold production of the Western states. There are over thirty placer gold districts and many lode districts as well. Since to cover all of them properly would take a book of its own, we will only include those that fit the intent of this work.

Gold deposits here have been worked since the earliest days of the Indians. In fact, it was the gold ornaments worn by the Indians and seen by an early explorer that caused the Spanish invasion of this region. When this wandering soldier returned to Mexico, his stories of gold influenced Coronado to form a large expedition to find the source of the gold. He never located the fabled Seven Cities of Cibola during his expedition of 1540, but the later Spanish settlers are known to have mined the deposits until the late 1600s, when a revolt by the Indians forced them to stop.

Modern production began in 1828 when gold was discovered in the Ortiz Mountains at what is known as Old Placers. New finds were still being made as late as 1908, when the discovery of the placer deposits at Sylvanite caused a small rush to the area. There is still a lot of gold out there due to the crudeness of the early equipment and the lack of water in most areas. So get out your detector and your dry washer and go find it!

Sylvanite District

The Sylvanite District is located on the western side of the southern portion of the Little Hatchet Mountains. To reach the area, go south on State Highway 146-81 from its junction with Interstate 10 about twenty miles from Hachita. Highway 9 crosses 81 here. The deposits are south of here in the gravels in the washes and gulches. The most productive area is between Cottonwood and Livermore Springs.

The first recorded production from this region was in 1880. The next and largest activity was in 1908. Sylvanite was never a large producing area, but there have been a few good-size nuggets found here. You'll get a little color and maybe even a nugget of your own.

White Signal District

In the southern portion of the Big Burro Mountains is located the White Signal District. To reach the area, go northeast from Lordsburg on Highway 90 about twenty-nine miles. There are several dirt roads leading into the various diggings.

The deposits here have been worked since at least the early 1880s. The most productive areas are in Gold Gulch and Thompson Canyon. There has also been some gold recovered at Gold Lake, ten miles east of Gold Gulch. You can get a little color in most of the washes of this region. There was quite a bit of activity here during the Depression years, as in other areas. The most likely source of the placer deposits is in the hard rock veins in the nearby mountains. It's a good area for dry washing and nugget shooting.

Hillsboro District

Also known as the Las Animas Placers, this district has been a good dry washing area. The deposits are on the southern and eastern portion of the Animas Hills. They can be reached by taking Highway 152 west from Highway 25. There are dirt roads heading north from Highway 152 (about five miles before the village of Hillsboro) that will take you into the actual diggings.

The most productive areas have been on the east side of the hills. Gold Run Gulch is the biggest producer, but the deposits in Snake Gulch and Wicks Gulch are also good. Try the area between the Rio Percha on the south and Dutch Gulch on the north and you should always get a little color for your efforts. The placer deposits were first discovered in 1877 and have been worked off-and-on since then.

Orogrande District

This district is located on the southern flank of the Jarilla Mountains in southern New Mexico, close to the Texas border. To reach this area, from Alamogordo, go south on Highway 70 to Highway 54, thirty-five miles to Orogrande. There are good dirt roads leading north to the deposits. It is about a mile-and-a-half to the diggings.

The placers here have been worked since the early 1900s. Most of the work has been small-scale, since the deposits are in caliche-cemented gravels and difficult to separate. This might be a good area to use your gold detector to locate the best spots and not have to dig a lot of empty prospect holes.

Other areas

There are numerous other areas where a dry washer and gold detector can be put to good use, such as Old Placers in the Ortiz Mountains, where mining has been going on since at least 1828. Another region worth investigating is the Pinos Altos Mountains, where modern-day sourdoughs are still finding enough gold to keep them going back until they hit the Big One. The Bayard region near Silver City (one of the most mineralized regions in New Mexico) is an interesting area, as most of the gulches here contain a little gold.

Treasure Tales

Some of the most famous treasure stories in America are located in New Mexico. Most are tales of lost mines or hidden caches of gold and silver. The tales begin as early as the 1600s and reach to the middle of the 20th Century. With all these stories, there must be one or two that could be rewarding—even today.

Victoria Peak

This tale begins in 1798, when a Padre LaRue, working with the Indians in Mexico, heard stories of riches to be found in mountains to the north of his location. When his wards faced hunger due to a crop failure, he decided to take his people north to find a better place to live. He took them to the area where he expected that gold could be found.

They settled in a basin which contained good water, below a lone peak that he had been told was rich in gold. This was Soledad Peak or Victoria Peak, as it was also known. It was not long before a rich vein of gold was discovered. The Padre and his Indians worked the mine for several years, using only as much gold as needed to buy supplies and storing the rest in a large cavern that was part of the mine.

During this time, his superiors became concerned when nothing was heard from the Padre and his flock. When messengers sent to his original location reported that he had moved his colony to the north without permission, troops were sent to find him and demand an explanation. When the Padre learned of the approaching soldiers, he ordered his men to hide the mine and all traces of their operation.

The soldiers had heard that the Indians had been buying supplies with gold and tried to force the Padre and his people to show them the source of the gold. When they refused, a fight broke out; the soldiers opened fire on the small band. Most were killed; those who escaped were never heard from again. Even those who were captured still refused to tell. After searching for the mine without success, the army returned to Mexico with their story.

In 1937, Doc Noss (Milton Ernest Noss), a resident of Hatch, New Mexico, said he had discovered the lost mine while hunting in the San Andreas Mountains. Noss said that he had a cave with a stack of gold bars in it on Soledad Peak. I saw a picture somewhere once of Noss holding a gold bar, but most reliable sources say there were never any gold bars seen.

According to the story, he accidentally dynamited the tunnel entrance closed. After that, he got investors to help him open the passage, but all they did was lose their money. During World War II, the area was taken over by the government and closed off. In 1949, Noss got into a shoot-out with a man named Ryan and lost. The truth is buried with him.

His ex-wife Ova tried to recover the treasure, but was caught trespassing on government land and was run off. John Dean mentioned this tale at the Watergate Hearings. In my opinion, if there was any gold hidden there, the government would have claimed it years ago.

Other lost treasure stories like the famous Lost Adams Diggings and the Lost Sublett Mine, as well as Apache leader Geronimo's mine, are located in New Mexico. For more details, pick up one of the excellent treasure books, like Tom Penield's *Guide to Treasure in New Mexico*, or *Lost Mines of the Great Southwest*, by Mitchell.

Utah Areas

Utah does not contain a large number of gold districts; only one, the Bingham District, produced a large amount of gold. Nevertheless, there is still gold being found, and maybe one of these days a really good discovery will be made.

There are just a couple of districts in the Great Salt Lake Desert Region; we will deal only with them. The other placer gold districts like the Colorado River, Green River and (of course) the Bingham, are worth looking into. You may not get rich, but you'll find a little color and even a nice nugget now and then. I guarantee that simply getting out will reward you.

House Range District

The House Range District is located 44 miles southwest of Delta, between Highway 6-50 and old Highway 50. The placer deposits are in Amasa Valley and Granite Canyon. There are dirt roads off both highways that will take you into the area. Note that Granite Canyon may be shown as Miller Canyon on some maps.

The first recorded production came in 1932 when the Depression sent people out in areas all over the west to try to make a little money prospecting for gold. The most productive area in this district has been the sand and gravel deposits of the Amasa Valley. Most of the gold from Granite Canyon has come from the gravels at the place where the east and north forks join the creek bed in the main canyon. The gold probably comes from quartz deposits on North Peak, but the geologists say they are not certain where the source of placer gold is located.

Detroit District

There have been reports of placer gold being recovered in canyons and washes in the Drum Mountains, which are located in Juab County. There are some lode mines here that have had a small amount of production; the veins they tap are probably the source of the placer deposits.

Treasure Tale

There is a story told of a Mexican sheepherder who worked in the House Mountains who found a lost Spanish mine. Sometime in the late 1930s a sheepherder brought a small amount of gold to a dentist in the town of Delta and offered to sell it to him. The dentist bought the gold and said he would buy more in the future if the shepherd found any more. He asked where the gold came from, but received no answer.

Several months later, the sheepherder returned to the dentist with a twenty-pound bag of gold—which was more than the dentist could use—so no sale was made. The man left and was never seen again. A rancher in the area said that one time the sheepherder showed him an old map with directions to a Spanish mine in the mountains. The Spaniards had been run off by the Indians and left everything behind.

Two different mines have been found with remains of Spanish equipment, but neither of them have turned out to be productive. There may be another mine still hidden out there, waiting to be found. Perhaps you will find it.

Southern Nevada

I love to prospect in this state because it has not been abused like so many other places. The remoteness of the mining districts, along with the harshness of the weather and the ruggedness of the terrain have kept most of the areas from being overworked. I've been prospecting in some of these areas and not seen another soul for days.

Nevada is our fifth-largest gold producing state. Mining is the second-largest industry in the state. The 6.7 million ounces of gold mined here in 1993 represented 70 percent of U.S. production and eleven percent of the world's production. Most of its gold has come from lode deposits and as a by-product of the silver and copper mines. There as been quite a bit of placer mining done here as well, most of it accomplished by dry washing.

As in most regions, it was miners working the placers who eventually located the lode deposits. It was "that damn blue stuff" that flustered the miners in the early days in Gold Canyon, by plugging up their rockers and sluice boxes, that would make some of them rich men later. The blue stuff turned out to be silver—in fact, some of the richest deposits ever found.

The earliest known gold placering in the state was in Tule Canyon sometime before 1848 by miners from Mexico. There is no record of how much gold they recovered, but some estimates place it as high as a million dollars.

We will deal only with the placer and lode gold deposits of the desert regions in this book. There are still better possibilities here for the prospector than other areas because of the remoteness of the mining districts along with the harshness of the weather and the ruggedness of the terrain have kept many less-hardy souls from overrunning the gold bearing regions.

Searchlight District

This is the highest-producing district in Clark County, with nearly a quarter-million ounces recorded. There have been reports of placer gold being found here, but all the recorded production has come from lode mines. Searchlight is thirty-six miles south of Boulder City on Highway 95.

Eldorado District

Most of the placer mining in this district has been done in the Eldorado Canyon area. There have been several shafts sunk in the canyon to try to reach pay dirt, so be careful out there. In addition, there are lode deposits to look for. To reach this district, take Highway 60 off Highway 95 south of Boulder City.

Gilbert District

Since this is also known as the Desert District, even though it is located on the side of a mountain, I figured we should include it in this book. There have been reports of small amounts of gold being found in this district, which is located in Esmeralda County and located on the northern slopes of the Monte Cristo Mountains.

Goldfield District

One of the last gold rushes in this country was to the Goldfield District. The first deposits were discovered here in 1902 and mining began in earnest the next year. There are lots of remains of the period still here and well worth a trip to see. All the production came from lode deposits, which in some cases were quite rich. Goldfield is located on Highway 95 about 25 miles south of Tonopah.

Gold Point District

This is a well-preserved gold camp near Goldfield. There are still a few people living here, so be careful when you go exploring. Small amounts of placer gold have been found here, but most of the production has come from lode deposits. The veins occur in calcareous shale near intrusive granite.

To reach Gold Point, go south from Goldfield about fifteen miles on Highway 95 to Highway 3, then go west on Highway 3 for seven more miles to Highway 71 (which is actually a gravel road). Follow this road seven miles more to reach Gold Point.

Klondyke District

This area in the southern Klondyke Hills, about fourteen miles south of Tonopah, has an interesting history. The Chinese are said to have worked the placers here in the 1870s. A large nugget worth $1,200 at the time was found here. This could be a good dry washing area.

Tule Canyon

One of the first gold placering areas worked in Nevada, recorded production began here in 1871—but miners from Mexico were working the deposits long before that. The placer deposits are spread over a ten-square-mile area. The most productive areas are in Tule Canyon and the gulches flowing into it.

The Tule Canyon District is located about ten miles south of the town of Lida at the southern end of the Silver Peak Range, between the Sylvania and Magruder Mountains. You can reach Lida by going nineteen miles on Highway 3 from Highway 95. It's another good nugget shooting and dry washing area.

Eagleville District

The Eagleville placers are located a few miles south of the old mining camp of Fairview, in a canyon south of the Eagleville Mine. The area is about halfway between Frenchman and Highway 23, south from Highway 50. Be careful, because nearby the military uses the land as a bombing and aerial gunnery range.

Lynn District

This district is located near the huge Carlin Mine operation; the placers are in Lynn, Sheep, Rodeo and Simon Creeks—all of which are dry most of the year. They can be dry washed or metal detected. The Lynn District is about eighteen to twenty miles northwest of the town of Carlin in the Tuscarora Range. Gold is found in the gravels at the upper end of the ravines; it is coarse and angular and of a very high purity.

Awakening District

Located in Humboldt County in the northern portion of Nevada, the Awakening District has seen some placer mining, but most of its gold production has come from load deposits. The gold has been found in the Slumbering Hills, near the ghost town of Davytown, about thirty miles north of Winnemucca.

Dutch Flat District

Some good-size nuggets have been found here, so you might try nugget shooting as well as dry washing. This is mainly a placer mining area, with about ten thousand ounces being reported. The deposits are located on the western slopes of the Hot Springs Mountains. To reach this area, take the good gravel road north out of Golconda to the dirt road that skirts the southern end of the mountains to the Dutch Flat Diggings.

Potosi District

As gold mining goes, Potosi is a fairly new area. The first recorded production did not occur until 1938. It is a lode mining area with some very rich veins. I suspect that you wouldn't mind finding a rich vein, instead of dry washing or nugget shooting, so it might be worth a look. Take Highway 18 north from Highway 80 about sixteen miles to a road going north. Continue about ten miles to the Getchell Mine. Search this area for interesting outcroppings.

Winnemucca District

There have been small amounts of placer gold found in the gulches around Winnemucca. All the recorded production of over 35,000 ounces has come from the lode deposits here.

Bullion District

This is a large district about thirty miles south of Highway 80, off of Highway 21. The best placering areas are in Triplett and Mill Gulches on the west side of Crescent Valley. There are some good lode deposits here as well.

Prospecting Tips

As long as you are out on the Utah and Nevada Deserts, keep your eyes open for other valuable minerals. These areas have produced silver, lead and copper, as well as gold. Some nice crystals have also been found here.

Geology of Placer Deposits

In order to find that golden dream you are seeking, you need to have some knowledge of placer deposits. A lot of our information comes from the early miners and prospectors who climbed, dug into and checked every mountain, canyon, stream, river, and creek. This is still the best method, as geologists admit that even today they don't know everything there is to know about the remaining gravel deposits.

In a *Mineral Information Service Bulletin* put out by the State of California Division of Mines and Geology, they stated "The geologic history and structure of the buried channels are so complex that the best of engineers have been baffled by them. Fragmentary benches and segments of rich gravel deposits which still rest in positions completely hidden from the surface, or even from the underground passages which enter into the lower main channels, afford alluring possibilities to the geologist as well as the prospector." They are telling us that there is still a lot of gold out there; you've as good a chance at finding it as any geologist.

The number one thing to keep in mind is that most areas have been prospected at one time or the other. Don't waste a lot of time in areas that have not proven to be productive in the past. Search the areas that are known to be gold bearing and take advantage of the knowledge gained by those who went before you.

There are several types of placer deposits which are classified as to how they were first formed. The basic placers are:

(1) Residual placers or "Seam Diggens."
(2) Eluvial or hillside placers, representing transitional creep from residual deposits to stream gravels.
(3) Bajada placers, a name given to a peculiar type of desert or dry placers.

(4) Stream placers, which have been sorted and resorted, and are simple and well merged.
(5) Glacial-stream placers, which are for the most part profitless.
(6) Eolian placers, or local concentrations caused by the removal of lighter materials by the wind.
(7) Marine or beach placers.

Of the seven types, the stream placers have been the source of most of the placer gold mined in California. Stream placers consist of sands and gravels sorted by the action of running water. If they have undergone several periods of erosion and have been resorted, the greater the concentration of heavier minerals.

Deposits by streams include those of both present and ancient times, whether they form well-defined channels or are left merely as benches. All bench placers, when first laid down, were stream placers similar to those of the present stream deposits. If not reworked by later erosion, they are left as terraces or benches on the sides of the valley cut by the present stream. These deposits are called bench gravels. In order to understand stream placers, streams themselves must be studied in regard to their habits, history and character.

Residual placers are formed when the gold is released from its source and the encasing material broken down. This is most effectively done by long continued surface weathering. Disintegration is accomplished by persistent and powerful geologic conditions which affect the mechanical breaking down of the rock and the chemical decay of the minerals. The surface of a gold-bearing ore body is enriched during this process of rock disintegration, because some of the softer and more soluble parts of the rock are carried away by erosion.

After gold is released from its bedrock encasement by rock decay and weathering, it begins to creep down the hillside and may be washed down rivulets and gullys and into stream beds. On its way down the hillside, the gold is sometimes concentrated in sufficient value to warrant mining. These deposits are classified as eluvial deposits.

It is a common fallacy of some prospectors to attribute the forming of some placer deposits to the action of glaciers. Since it is the habit of glaciers to scrape off loose soil and debris but not to sort it, ice is ineffective in the concentration of metals. The streams issuing from the melting ice may sometimes be effective enough in sorting to create a deposit.

Bajada is a Spanish word for "slope," and is used to identify a confluent alluvial fan along the base of a mountain range. The total production of gold from bajada placers is small compared to other placer workings, due to the less efficient dry washing methods used in the past. The forming of a bajada placer is basically similar to a stream placer, except it is conditioned by the climate and topography of the arid region in which it occurs. The bulk of the gold that has been released from its matrix as it travels from the lode outcrop to the bajada slope is deposited on the slope close to the mountain range. The gold is dropped along the lag line—the contact of the basin fill with the bedrock. Most Eolian placers of the desert result from a bajada being enriched on the surface by wind action on the lighter materials.

Although the heaviest concentration of gold will be on bedrock, bulk concentration does not occur as in a stream deposit. Since a certain percentage of gold is still locked in its matrix, there is a strong tendency for less gold to reach bedrock and for more to remain distributed throughout the deposit than in the case of stream gravels.

There have been several beach placers found and worked along the Pacific Coast. Beach placers are a result of the shore currents and wave action on the materials broken down from the sea cliffs or washed into the sea by streams. There are two types of beach placers: present beaches and ancient beaches. Most of the gold-bearing beaches are found in northern California. Not a great deal of production has come from these deposits. Most of the gold will come from the rocks that are being eroded by the waves.

Placer deposits are preserved in several different ways. Since streams are constantly changing their position, fragments of their deposits are left isolated; for example, the beaches and terraces that are left at different intervals when a stream is cutting a deeper channel. These deposits that are left will eventually be eroded away unless something protects them. Burial is the most effective way a placer may be preserved. Mostly when the name "buried channel" is given to a placer, it is one where a stream has been covered by lavas, mudflows or ash falls. There are other ways by which they may be buried, such as:

(1) By covering with landslide material.
(2) By covering with gravel.
(3) By covering with lake deposits.
(4) By covering with gravel when the stream is choked.
(5) By covering with gravel when the stream course is lowered below the general base level of erosion.
(6) By the covering of older stream courses with alluvial fan material as conditions favorable to stream existence fail.
(7) By covering with glacial till.
(8) By covering of beach placers with marine sediments as the coast is submerged and elevated.

The gravel content of a placer may become firmly cemented when it is infiltrated by mineral matter, such as lime and iron carbonate or silica. The older the placer the more likely this is to occur. These cemented gravels sometimes are very hard to break down, which is why some tailings of old mines are profitable to work. The cemented gravels sometimes were never completely broken down as they traveled through the sluice boxes, and the gold was redeposited in the present stream bed.

The gold found in placers originally came from veins and other mineralized zones in bedrock when the gold was released from its rock matrix by weathering and disintegration. Many times the source of the gold in a placer would not be a deposit that could be mined at a profit, but the richer deposits usually indicate a comparatively rich source. Sometimes a rich placer will be developed when several low grade veins feed it over a very long time. The richest placers are created when there is reconcentration from older gold bearing gravels. For the most part, the original source of the gold is not far from the place where it was first deposited after being carried by running water.

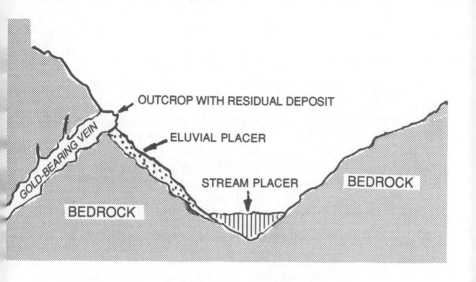

Cross-section showing transitional stages in the development of placer deposits.

Mineral grains—"black sand" to prospectors—that are very heavy and resistant to mechanical and chemical destruction will be found with gold in placer deposits. Black sand is generally principally magnetite, but other minerals you will find in your sluice box are garnet, zircon, hematite, pyrite (fool's gold), chromite, platinum, cinnabar, tungsten, titanium and even diamonds. You'll find a lot of other things, such as quicksilver, metallic copper, amalgam, nails, buckshot, BB's, bullets, and what-have-you.

AVERAGE SPECIFIC GRAVITY OF SOME MINERALS

Mineral	Specific Gravity
Mica	2.3
Feldspar	2.5
Quartz	2.7
Hornblende	3.2
Garnet	3.5
Corundum	4.0
Magnetite	5.2
Silver	7.5
Gold	19.2

The very high specific gravity of gold, six or seven times that of quartz (increasing to nine times under water) is what causes the gold to work its way to bedrock or false bedrock, or any point that it can go no further. Once it is trapped on bedrock, the stream has great difficulty picking it up again. The specific gravity of gold is 19, that is, it weighs 19 times as much as an equal amount of water to its mass.

Due to their insolubility, the finest particles of gold are preserved. A piece of gold worth less than a dollar can easily be recognized in a pan. Since gold is so malleable, it will be hammered into different shapes by stones as they tumble along in the stream. It will not be welded together to form large nuggets as some people believe. Particles of gold may be broken down, however, from another piece.

Geologists have shown that the largest masses of gold come from lodes and not placers. The more rounded and flattened nuggets have probably been in the stream for a longer amount of time and have taken more knocking around than the ones that show the original crystalline form. The crystalline nuggets are known as coarse gold and probably have not travelled as far from the source in the free state.

The gold found in the more ancient placers has a higher degree of fineness than gold whose source is nearby. This may be due to the removal of alloyed silver by the dissolving action of the water.

Accumulation of gold in an important placer deposit is not just pure coincidence, but is the result of some fortunate circumstances. In areas where nature has provided extensive mineralization, rapid rock decay and well-developed stream patterns, there is an opportunity for gold placers to be formed.

Basically, what happens is fairly simple. In areas where gold has been deposited, the power of the stream has become insufficient to carry off the particles of gold that have settled. How rich the deposit is will depend on how complete the loss of transporting power is, as well as the ability of the bedrock to hold the deposited gold, plus the general relationship of the gold sources to the stream. When a stream is eroding, the materials in reach of its activity are constantly moved downstream.

Cross-section of a river that has suffered down-faulting on the upstream side, creating a pocket to trap the gold, gravel, sand and silt.

During this movement, a constant sorting is taking place, which causes a concentration of the heavier particles. Deposition then takes place in the stream when the velocity is decreased, either by changes in volume or grade. When this happens, the gold is laid down with other sediments. Sometimes the placer gold may be trapped in irregularities in the bedrock, without considerable detrital material being trapped with it, but extensive placers (as a rule) are not formed by irregularities in the bedrock alone.

When the bed of the stream is the actual rock floor of the valley, this is true bedrock. When the gravels become covered with volcanic or other materials, the stream will flow over this new floor, making deposits on what is known as false bedrock. Thus, an area may contain two or more layers of gold bearing gravels. An easy way to see how a stream lays down these various layers is to study areas where road cuts have exposed ancient stream deposits, also in canyons where benches can be seen.

A smooth hard bedrock is a very poor location to develop a good placer deposit. The bedrock formations that are highly decomposed, with cracks and crevices are good, and those of a clayey or schistose nature are rated excellent in their ability to trap gold.

Here are some figures from a report by the California Division of Mines and Geology on the size of material carried by a stream flowing at different velocities:

3 in. per sec.	0.170 mph	will just begin to work on fine clay.
6 in. per sec.	0.340 mph	will lift fine sand.
8 in. per sec.	0.454 mph	will lift sand as coarse as linseed.
10 in. per sec.	0.5 mph	will lift gravel the size of peas.
12 in. per sec.	0.6819 mph	will sweep along gravel the size of beans.
24 in. per sec.	1.3638 mph	will roll along rounded pebbles 1 inch in diameter.
3 ft. per sec.	2.045 mph	will sweep along slippery angular stones the size of hens' eggs.

As far as grade is concerned, a grade ranging from 30 to approximately 100 feet per mile will favor deposition of gold. Anywhere that the grade is greater than that, such as in mountain streams or in narrow canyons, will not be a good source of placer deposits.

When a stream leaves its mountain canyons and enters a more level country or a still body of water, the materials carried by the stream are deposited in the form of a fan or a delta. At the apex of this fan or delta the fine gold will be deposited and may never reach bedrock. The stirring action that occurs in the rugged mountains during times of floods, which permits gold to reach bedrock, does not take place in the delta.

To sum up, remember that gold is heavier than other material in the stream bed, and it will drop anywhere the flow or grade changes, causing the stream to slow down or lose its carrying power. These are the places you want to search. Each rainy season will bring new gold down from the hillsides into the stream bed.

Keep in mind that early miners were working deposits that developed over thousands of years. Try to find material that has not been worked before or try to reach hard-to-get-to areas where the gravels have not been worked as much. The most important thing to keep in mind is that "GOLD IS THERE TO BE FOUND."

Some Tips on Desert Prospecting

The deserts of California contain some of the largest reserves of placer gold to be recovered in the state. Olaf P. Jenkins, Chief Geologist for the State of California, in a series of articles for their monthly bulletin, listed the most likely places this gold would be. They were as follows:

1. Bajada Placers, or desert alluvial fan deposits, where gold is derived directly from the original mineralized bedrock source.
2. Desert Placers, where the gold is reconcentrated from more ancient gold bearing gravels.
3. Buried desert stream placers.

He also said, "In the Mojave Desert, tertiary gold bearing streams may have once flowed south from the region which is now the Sierra Nevada. The problem of the 'dry placer' is one of considerable importance, and when more is learned about it, an immense gold wealth may be discovered which has not yet been glimpsed."

That's pretty heady stuff from the State Department of Mines and Geology which is, as a rule, very conservative in its estimates. Now, let's see what you'll need to get that gold for your own.

Water

Water is as important to a prospector as anything else he uses. You should carry at least two five-gallon cans of water on every trip, one for emergency use for drinking or the vehicle and the other for panning. You will need a small wash tub to recycle the water. When you find a place you want to work, just pour the water from the five-gallon can into the wash tub and do your panning. When you are finished, let the water clear and return it to the can.

Dry Washer

You can buy a dry washer ready-made or build your own. I have a Keene Vibrostatic Concentrator which, for my money, is the best portable dry washer made. The concentrator has a high-static fan to force air up into a special plastic tray, where it gains an electric charge from the plastic.

The air then moves under pressure through a special artificial fabric where the charge is increased. Material is shoveled into the concentrator through a screen which automatically classifies the material, letting only gravel less than half-an-inch wide into the concentrator. The material then works its way down to the plastic tray.

Gold is non-magnetic, but it has an affinity for an electrostatic charge and it will be attracted to the special cloth. The concentrator also works like a regular dry washer and traps the gold behind riffles. I have been able to save even very fine pieces of gold with this machine.

Dry washing goes back to the earliest days of working desert placers. Since there was no water to separate the gold from other materials, miners devised several methods using a flow of air to concentrate. In the past, most dry concentration was slow and inefficient.

Even today, most dry washers will have trouble recovering gold from moist sand and gravel after the top layer of dry sand has been removed from the area being prospected. Always make it a practice to run your material through the dry washer more than once.

One of the earliest methods of dry washing was known as winnowing. In winnowing, the coarse gravels are screened out and thrown away, then the fines are placed on a blanket. The blanket is picked up by the corners and the fines tossed into the air in a strong wind. The lighter material is carried off by the wind and the heavier minerals fall back into the blanket. The weave of the blanket also helps to trap the flour gold.

You can also dry pan the gravels, but unless you are awfully good, a lot of the gold will be lost. Currently the most popular method of working desert placers is with a bellows-type dry washer (see illustration). Gravel is shoveled onto a screen with a hopper underneath. The large coarse gravels (normally a half-inch or more) are separated from the finer material by dropping off the lower end of the screen, while the smaller material falls down into the hopper.

The smaller material is then funneled down into the riffle tray, where it flows over the riffles where the gold is actually trapped. The flow of the material is aided by air pushed upward by the bellows. The bellows can be operated either by a hand crank, small gasoline engine, or a 12-volt powered motor.

If you're a one-man operation, you'd do well to purchase a motor-driven concentrator so that you won't have to stop and crank the bellows every time you put a few shovels full of dirt into the hopper. A vibrostatic concentrator can keep two men busy shoveling and keeping the screen clear.

Working the Keene hand crank dry washer.

White Adhesive Tape

A roll of white adhesive is a must. No, it's not for doctoring; it's for marking your ore samples. Whenever you locate a vein that you feel might be valuable and you think should be assayed, use the tape to mark the location you got it from. God only knows how many pieces of rich ore I've seen that no one knows how or where they came from.

Write the date and exact location that you found it on a piece of the tape and stick it on each piece of ore you bring in from the field as a sample for assaying.

Snake Bite Kit

You shouldn't begin a desert trip without bringing a snake bite kit along for protection. I know of several prospectors who owe their lives to the time given them by a snake bite kit. I've never been bitten, but I've come close.

One hot summer day when working in an area south of the Turtle Mountains, I found myself running low on water. I decided to go to the Colorado River to replenish my supply. I made my way to Highway 95, which runs alongside the river for about forty miles after it leaves Blythe on its way to Needles. After I got to Highway 95, I followed it for a few miles to a spot that looked pretty close to the river.

I parked my Bronco and took a couple of water cans and headed for the water. The brush gets pretty thick as you get near the river and I had to raise my foot pretty high to push it down to get through. I was in the middle of this brush when I ran into a four-foot rattlesnake. I almost stepped on him as I was pressing down the brush with my foot. For a moment, time seemed to stand still as I tried to stop my foot from coming down on his back. That snake looked at me as if I was crazy, then we both took off in opposite directions. If I live to a hundred and five, I'll never forget the look in that snake's eyes. Some of the brush must have hit him and scared him for the snake to take off that fast.

When people get bitten, it's mostly on the arms or hands. Be careful when you're climbing and don't stick your hands in holes without first probing with a long stick. One last note: if bitten, head for the nearest hospital or doctor's office as quickly as possible. Pack the wound and the area around the wound with ice and leave it packed until you obtain medical aid. The ice will help retard the spread of venom.

Compass

Pick up a good compass before you go; it may save your life. Hundreds of stories are told of people losing their lives wandering aimlessly only a few miles from a main highway. Select a prominent landmark and note its direction from your

Gold Scale. (Courtesy of Keene Engineering)

campsite. Note also what direction the nearest main road lies.

Metal Detector

A metal detector is not a requirement—and some old timers look down their noses at them—but the truth is that recent developments in metal detector technology have really made an impact on gold recovery in desert regions. Nugget shooting is the fastest growing segment of prospecting today.

Detectors such as the Gold Bug, the Gold Master II, Mine Lab Gold Striker, and others that are designed strictly for hunting gold nuggets are enabling people with little or no prospecting experience to recover more gold in a shorter period of time than many an old hand ever did. I've used all three of these detectors with some success, but not without a little frustration as well.

I guess there is something to be said in getting gold every time you dig, but it will never replace dry washing or hard rock mining for recovering large amounts of gold. I've seen a lot of gold nuggets that have been found nugget shooting, so the detectors do work. It's a slow process and there are some tricks to it, but you may want to try it yourself.

Most detectors work in an "all metals" mode, and there seems to be no place you can go that doesn't have some trash such as lead from bullets. Even the salts in the skin of your hand may cause a false signal when you pass your hand over the coil with the dirt you have dug, so get a long-handled plastic measuring cup to check the dirt when you dig for a signal. In fact, ground balancing is a slow process and must be checked often if you move around a great deal in searching an area. Mine Lab has developed a self-ground balancing unit (and others are working on it), so balancing may not be a problem in the future.

I told one manufacturer that I must be growing up, because I threw my shovel over the cliff—and not the detector—the first time I field-tested one of these new detectors. I always take at least a standard detector with me, whenever prospecting an old area that had a camp, to search for old coins and artifacts just to break up the day.

You can prospect to some degree with a standard metal detector by using the "mineral" mode to locate hidden black sand

*Probing and testing with a metal detector.
(Courtesy Fisher Research Laboratory)*

deposits. Clean the ground off down to bedrock if you can. Now put the detector in the "all metal" mode and tune it as fine as possible. Pass your detector over the area as slowly as possible and watch your meter needle carefully. If you have headphones, use them, too. Check out every signal you get. Some prospectors I know get good results using this method of nugget shooting in desert washes.

Gold found in placer deposits in the desert, like all placer gold, came from veins through weathering and disintegration of the matrix holding the ore. A rich placer deposit is not always a sign of rich veins nearby, as several lowgrade veins may contribute to a rich placer deposit, but normally the richer placer deposits are a good indication of a rich source. Even richer deposits may come from the reconcentration of older gold-bearing gravels.

Mainly, though, the original source of gold in the placer is not far from its first place of deposition. The weight of the gold, nearly seven times that of quartz (and its malleability) are the primary factors in the creation of placer deposits.

Bajada Placers are found in gravel deposits along the lower slopes of the mountain ranges in the desert. The richest deposits are along the lag line on or near bedrock. Because concentration does not occur on the bajada slopes as it does in stream beds, and all the gold has not separated from the matrix, there is a tendency for less gold to reach bedrock than in stream gravels.

Desert placers found in washes and streams are made of deposits from ancient gold bearing channels cut by the flow of the water and wind and from new gold washed down from exposed veins. If you locate any gravel benches on canyon walls, always test them in several places along the old bedrock.

Another source of placer gold is dry lake beds. Try to locate the places where ancient streams fed the lake. Any place where you can reach bedrock is also a potential source of desert placer gold.

There are places in the desert where volcanic rock has covered ancient stream beds or picked up the gravels and mixed them with lava. Many times these mixtures have contained vast amounts of gold and may be the source of the black gold told of in so many of the lost treasure tales.

In prospecting for lode deposits, keep in mind the type of

ore where the gold is known to occur in the various districts listed in this book; this will save you a lot of time and wasted effort. This is not to say that it would be impossible for gold to be found in any other matrix, but it is improbable.

Look for float (pieces of the vein that have broken off and worked their way down a hillside or stream bed) and follow it to its source. Take as much ore from the vein as you can carry for assaying. The more ore that is assayed, the truer the reading on the real value of the outcrop will be.

Learn to identify quartz by buying a few samples at a lapidary store with a mineral identification set. Any quartz outcropping that is iron-stained or contains fine-grain pyrite should be assayed.

Any small rounded hills in a rugged area should be checked for mineralization. Brush growing on canyon walls in an otherwise barren area is also a sign of mineralization; carefully study these locations.

Don't be afraid to use your shovel and dig down in gold-bearing regions, as many gold bearing veins have never reached the surface. The gold is out there, billions of dollars worth. Someone is going to find it someday. It could be YOU.

Panning in the Desert

Most prospectors prefer to pan their material in water, because dry panning is very slow and the recovery rate is not as good as in water. So, like most prospectors working in the desert, I save my concentrates from dry washing and sampling and take them to the nearest water source for washing.

Some desert prospectors carry a large wash tub and a supply of water with them and and pan right on the spot. Some even devise ways to circulate the water, filtering it to keep the sediments under control, so that the water being used for panning in the tub isn't clouded.

Remember that many watercourses in the desert have water running in them during the late winter and early spring, so you might want to schedule your explorations to take advantage of that situation. Besides, the temperatures are more pleasant then. All in all, it is better to find the nearest stream and pan your concentrates there, if you can.

How to Pan

There are several types of pans available, including steel, plastic and copper. I recommend the medium-size black plastic pan for beginners (see illustration). It is light in weight and has molded riffles in the side to trap flakes of gold. Learning to pan is not difficult, and with a little practice, you'll be panning like a pro.

1. Select a spot in the stream where the water is sufficiently deep to submerge your pan completely, and it is flowing fast enough to carry off the muddied water.

2. Fill your pan one-half to three-fourths full of the dirt or gravel you wish to wash. Dip the pan into the stream and fill it with water. Holding the pan with one hand, use the other hand to break up any clods that may be in your material.

3. With a gentle circular motion, swirl the gravel around in the pan. This will cause the dust and clay to come to the top.

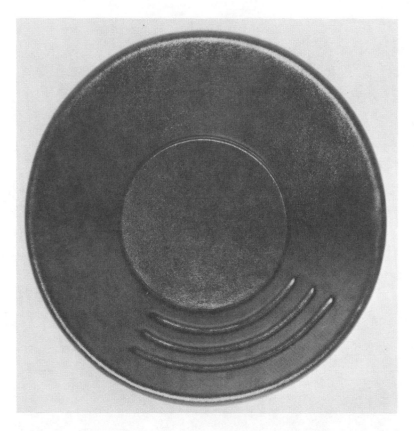

The plastic pan; note the built in cheater riffles on the side of the pan.

4. Submerge the pan again and continue the circular motion, allowing the lighter material to float away.

5. When the water has cleared, pick out the largest pebbles and throw them away.

6. Keeping the pan just under the water, tip it slightly away from you. Begin swirling the water around the pan and with a slight, forward, tossing motion, carefully forcing the lighter materials out of the pan. When using the plastic pan, be sure that the cheater riffles are toward the front, so that the gravel is passing over them as you toss the gravel out.

"Who says prospecting is only a man's game"
(Photo courtesy of Keene Engineering)

7. From time to time, level the pan and shake it back and forth. This will cause the lighter materials to come to the top and the gold to sink to the bottom.

8. Repeat the above process until only the heavier material remains. This will usually be black sand with the gold underneath. These are your concentrates or black sand.

9. Take your pan out of the stream and pour off most of the water, leaving enough to cover the black sand by about half an inch. Shake everything to the front of the pan by tilting it forward. Now tilt the pan toward you and start the whirling again until the black sands are removed and only the gold remains. Use your tweezers to pick out the flakes and put them in your specimen bottle for safekeeping.

When using the plastic pan, remember that black sand is usually magnetic, but your pan is not. Use a magnet to help you in the final processing. Place the magnet on the underside of the pan and tilt the pan slightly while moving the magnet in a circular motion. This will separate the magnetic black sand from the gold.

Recording Requested By:

Name _____

Address _____

_____ For Recorder's Use

NOTICE OF LOCATION - PLACER CLAIM

To whom it may concern:
1. Name of claim _____
2. Date of location: (date location notice was posted on the claim)
_____ (month/day/year)
3. Description of the discovery monument is as follows: _____
_____
5. This placer mining claim is located on land in the U.S. Public Land Survey.
The description of the claim by legal subdivision is as follows: Acres _____ Section _____
Township _____ Range _____ Meridian _____ A.P. _____
Mining District _____ County of _____ State of _____
If claim is not in the U.S. Public Land Survey, the boundaries of
the claim are described as follows:

The discovery monument is located at the point of discovery about:
_____
_____

Commencing at the discovery monument where this notice is posted, thence _____ to the
_____ corner which is the beginning point to describe the boundaries; thence _____
feet to the _____ corner, thence _____ feet to the _____ corner,
thence _____ feet to the _____ corner, thence _____ feet to the _____
corner, thence _____ feet to the point of beginning.
6. The undersigned locator(s) have defined the boundaries of the claim by erecting at each corner, or nearest accessible points a conspicuous and substantial monument. Each corner monument contains markings to designate the corner and name of the claim.

Signatures of Locator(s)

_____ Date _____   _____ Date _____

_____ Date _____   _____ Date _____

_____ Date _____   _____ Date _____

Name_____ Address_____

Name_____ Address_____

Name_____ Address_____

Name_____ Address_____

Name_____ Address_____

Name_____ Address_____

How to Stake a Claim

If, by chance or skill, you should strike El Dorado and locate a rich deposit of gold, you will need to know how to claim it for your own. Remember that the first men involved in writing mining law were either miners or men connected in some way with the Gold Rush. Men like Hearst, Armour and Stanford, whom we associate with newspapers, meat packing and universities, began their fortunes in the gold fields. It was these same men who helped draft California's laws.

In some states, land laws still favor the miner. If you own your home, you probably are aware that you do not always get mineral rights with your land title. Mineral rights have to be claimed and proved, if contested. Mineral rights belong to the locator.

If you locate a deposit you feel is worth developing, there are some basic things you can do to protect your discovery. First, as you know, there are two types of gold mines: placer and lode. Each has its own set of rules and regulations.

A placer claim may consist of as much as twenty acres for each person signing the location notice. Placer gold, again, is free gold found in gravels and alluvial deposits. You must post a Notice of Location on a post, tree, rock in place, or on a rock monument you build, showing the name of the claim, who is locating it, the date and the amount of area claimed. You must mark the boundaries of your claim so that it may be easily traced. You must also identify the location of your claim by reference to some local landmark or natural object (such as a stream or rock formation) or a permanent monument. Finally, you must file a copy of the Notice of Location with the local county recorder within ninety days of the date of location.

A lode claim uses a different procedure. Lode gold, again, is gold in a vein, which must be mined and separated from its mother rock. The claim may consist of as much as 1,500 feet along the sidelines of the vein, and three hundred feet on each side of the middle of the vein.

Within ninety days of the location of any lode-mining claim, you must place a post or stone monument at each corner of the claim. The posts must be at least four inches in diameter and the stone monuments must be at least eighteen inches high. A Notice of Location form must be posted on the claim. Also, in most states, a Notice must be filed with the county recorder and the state office of the Bureau of Land Management within ninety days.

This chapter is not intended to be a complete thesis on all the rules and regulations regarding mining claims, but it will help to protect you in the initial phases should you make a rich discovery. For more details, you may be interested in *Stake Your Claim,* by Mark Silva, which explains the legal and business aspects of gold prospecting and claims.

Assaying and Refining

If you think you have found a good gold spot, you'll want to determine its real value. Take your samples to an assayer. He will give you a full report on your sample. Who knows, you might have also located a silver mine as well!

If your assay shows a good percentage of gold, you have several options. You can contact a mineral resource company to help you develop the lode, or you can develop the mine yourself. (Incidentally, a weekend prospector should not casually tackle amalgamation. Mercury is a very dangerous material, and it should be used only by those well-trained in handling it.)

In the latter case, you'll ship the ore to a smelter to be refined and returned to you. The refinery may buy it directly from you as well.

If you're successful at placering, you may be able to sell your gold directly to a refinery, jeweler, dentist, assayer or prospector's supply store. Check the newspaper for the current price of gold before approaching a possible buyer. Your unrefined gold will sell for less, naturally, than the price of refined gold quoted in the paper; however, the daily market price will give you a ball-park figure to start from. Keep in mind that any fair-size nugget will be worth more than its weight, because nuggets are collectibles in themselves and are sought after by collectors.

Now you know where to look, how to find gold and what to do with it. So come on out and join us. I'll be looking for you!

RECORDING REQUESTED BY

AND WHEN RECORDED MAIL TO

NAME
STREET
ADDRESS
CITY &
STATE

(SPACE ABOVE THIS LINE FOR RECORDER'S USE)

LODE MINING CLAIM LOCATION NOTICE

Notice is hereby given that the undersigned have this _____ day of _____, 19____ located a lode mining claim on public (surveyed) (unsurveyed) lands in _____
Mining District, County of _____, State of California.

1. The name of this claim is _____. It is situated in:

NE¼ ☐	NW¼ ☐	SW¼ ☐	SE¼ ☐	Sec. _____ T. _____ R. _____ Mer. _____
NE¼ ☐	NW¼ ☐	SW¼ ☐	SE¼ ☐	Sec. _____ T. _____ R. _____ Mer. _____
NE¼ ☐	NW¼ ☐	SW¼ ☐	SE¼ ☐	Sec. _____ T. _____ R. _____ Mer. _____
NE¼ ☐	NW¼ ☐	SW¼ ☐	SE¼ ☐	Sec. _____ T. _____ R. _____ Mer. _____

2. The locator or locators of this claim are:

   Name(s)   Current Mailing or Residence Address

3. The number of linear feet claimed in length along the course of the vein each way from the point of discovery is _____ (not to exceed 1500 feet) in a _____ direction, and _____ feet in a _____ direction. The width on each side of the center of the claim is _____ feet. (not to exceed 300 feet)

4. The claim, described by reference to some natural object or permanent monument as will identify the claim located, is as follows: _____

5. This claim covers, among other things, all dips, variations, spurs, angles and all veins, ledges, and other valuable deposits within the lines of this claim, together with all water and timber and any other rights appurtenant, as allowed by the laws of this State and/or the United States.

Dated: _____, 19____   Signature(s) _____

NOTICE IS HEREBY GIVEN by the undersigned locator(s) that in accordance with the requirements of California Public Resources Code:

1. The above notice of location is a true copy of said notice; and is hereby incorporated by reference herein and made a part hereof.

2. The locator(s), within the prescribed time, as required by law, have defined the boundaries of this claim by erecting at each corner of the claim and at the center of each end line, or nearest accessible points thereof, a conspicuous and substantial monument; and each corner monument so erected bears or contains markings sufficient to appropriately designate the corner of the mining claim to which it pertains and the name of the claim. The date of marking is: _____; and the description of monuments is: _____

3. The United States survey within which all or any part of the claim is located is:
   Section _____, Township _____, Range _____, _____ Meridian.

Dated: _____, 19____   Signature(s) _____

CLAIMS LOCATED AFTER OCTOBER 21, 1976, MUST BE RECORDED WITH THE BUREAU OF LAND MANAGEMENT **WITHIN 90 DAYS AFTER DATE OF LOCATION.**

(SEE REVERSE SIDE)

CSO 3800-2
6/79

WOLCOTTS FORM 1134—LODE MINING CLAIM LOCATION NOTICE—Rev. 3-80 (price class 3)

Sample of Lode Mining Claim Location Notice

Glossary

ALLOY. A solid solution of two or more minerals.

ALLUVIAL. Loose gravel, soil, or mud, deposited by water.

ARRASTRE. A circle of stones where ore was crushed during the early days of gold mining; a primitive but effective method of separating gold from quartz.

ASSAY. To evaluate the quantity and quality of the minerals in an ore.

BAR. Sand or gravel accumulated in rivers where the flow of water slowed down and lost its carrying capacity.

DIGGINGS. A claim or place being worked.

DIORITE. A granular, crystalline, igneous rock in which gold sometimes occurs.

DREDGING. A method of vacuuming gold-bearing gravels from river or stream bottoms.

DRY WASHER. A machine which separates gold from gravels by the flow of forced air.

GLORY HOLE. A small but very rich deposit of gold ore.

GRAVEL BENCHES. Gravel deposits left on canyon walls through stream erosion.

HARDROCK MINING.	Another term for lode mining.
HEADFRAME.	The support structure located at the entrance of a mine over a shaft. Used for hoisting.
HYDRAULIC MINING.	A very destructive and now outlawed form of gold mining used in the Gold Rush. Giant hoses were used to force great streams of water onto canyon walls containing gold-bearing gravels. The walls were washed away into sluice boxes, where the gold was then picked out.
IRON PYRITE.	A common mineral consisting of iron disulfide which has a pale, brass-yellow color and a brilliant metallic luster. Also called fool's gold.
LODE.	A vein of gold mined either through a tunnel or a shaft.
MATRIX.	The material in which the gold is found.
PLACER.	Free-occurring gold which is usually found in stream and river gravels.
POCKET.	A rich deposit of gold occurring in a vein or in gravels.
POKE.	A leather pouch used by old-time miners to hold their gold.
QUARTZ.	A common mineral, consisting of silicon dioxide, that often contains gold or silver.
RETORT.	A device used to separate gold from mercury.

RICH FLOAT.	Gold-bearing rocks worked loose from a lode.
ROCKER.	A device used by the early miners during the Gold Rush. This was a sluice box mounted on rockers with a hopper on the top to classify the material. Gravels were shoveled into the hopper; then water was poured on top, washing the gold-bearing material down over the riffles while the hopper was rocked. The rocking helped the gold to settle.
SCHIST.	A crystalline rock which is easily split apart.
SLUICE BOX.	A trough with obstructions to trap gold used in continuously moving water.
STAMP MILL.	A machine used to crush ore.
SULFIDE.	A compound of sulfur and any other metal.
TAILINGS.	The material thrown out when ore is processed. The tailings from the early mines, where the miners were sometimes very careless, have produced significant amounts of gold and other valuable minerals.
TERRACE DEPOSITS.	Gravel benches high on canyon walls.
WALL.	The rock on either side or a vein.
WIRE GOLD.	Gold thinly laced through rock.

Important Note

As this new edition goes to press, the California Desert Protection Act has just passed in Congress.

Due to some possible lawsuits and the vagueness of the jurisdiction of the Bureau of Land Management and the National Park Service over this vast area, what effect this will have on the areas in California listed in this book is unknown.

The BLM and National Park Service have very different mandates in their management of the public lands. Generally, the BLM mandate is multiple use and sustained use. You cannot collect or mine materials, however, on land managed by the National Park Service.

You should look for posted signs before entering an area or check with the nearest BLM office on land use and boundaries.

You can still collect on private land, but as always, be sure to obtain permission of the land owner before entering.

James Klein was born in Beechgrove, Indiana. In 1932, at the age of 14, he arrived in Los Angeles. He attended Los Angeles City College and UCLA, then went to work for the old *Daily News*. He later moved to the Los Angeles *Mirror-News*, and then on to the Los Angeles *Herald-Examiner*. For the next five years he was sales manager for B.S.I., an automotive battery company.

During his years as a newspaperman, he began doing small parts in movies and television shows. Introduced to gold prospecting by a friend, he caught gold fever almost at once. He now devotes his full time to acting and prospecting. He is a frequent guest on television talk shows, and has been seen in *Coming Home, Comes a Horseman, The China Syndrome, The Electric Horseman*, and *Tom Horn*.